SpringerBriefs in Physics

W0235287

JF Chen · Heejeong Jeong
MMT Loy · Shengwang Du

Optical Precursors

From Classical Waves to Single Photons

 Springer

JF Chen
Department of Physics
East China Normal University
Shanghai
People's Republic of China

Heejeong Jeong
Frontier Research Laboratory
Samsung Advanced Institute of Technology
Yong-in
Republic of Korea (South Korea)

MMT Loy
Shengwang Du
Department of Physics
Hong Kong University of Science
  and Technology
Hong Kong
Hong Kong SAR

ISSN 2191-5423   ISSN 2191-5431   (electronic)
ISBN 978-981-4451-93-2   ISBN 978-981-4451-94-9   (eBook)
DOI 10.1007/978-981-4451-94-9
Springer Singapore Heidelberg New York Dordrecht London

Library of Congress Control Number: 2013942656

Printed on acid-free paper

Springer is part of Springer Science+Business Media (www.springer.com)

Contents

Chapter 1
Introduction

Abstract The study of precursor could be traced back to 100 years ago, the time when Sommerfeld and Brillouin attempted to explore the propagation speed of a finite pulse. Precursors, generated from sharp rising edges of an optical pulse, therefore verify the speed limit raised by Einstein. As they pointed out, the information velocity never exceed the speed of light in vacuum c. The problem will become much more interesting when single photon source is involved. In this chapter, we would like to introduce the optical precursor in classical wave domain, and extend the discussion to the single-photon domain.

1.1 The Speed Limit of Light and Causality Principle

The invariance of speed of light is one of the founding bases of modern physics. In 1905, when Einstein published "on the Electrodynamics of moving bodies", he raised the following postulate as one of the two core assumption of special relativity [1]: the light is always propagating in the empty space with a definite velocity c which is independent of the state of the motion of the emitting body. Therefore, the speed of light in vacuum is c in any reference frame. The constant speed of light is so surprising, because the postulate changes the physical picture concerning our time–space. The time and space mingle together through the constant speed of light. According to the special relativity theory, the speed of light in vacuum c is the maximum speed that any object could be accelerated to reach.

However, the problem is that Einstein didn't specify the velocity of light pulse that the speed limit applied to. It is natural that there is only a single velocity of light pulse when it propagates in the vacuum or air due to the absence of the medium dispersion. Studies on the dispersive medium, especially on superluminal medium, give birth to a term called "fast light" or "superluminal velocities". Early observations on fast light occurred at the time when laser was invented. Basov et al. [2] investigated the propagation of a pulse through a collection of inverted atoms, which amplify the incident laser pulse. In this case, the intensity of

JF Chen et al., *Optical Precursors*, SpringerBriefs in Physics,
DOI: 10.1007/978-981-4451-94-9_1, © The Author(s) 2013

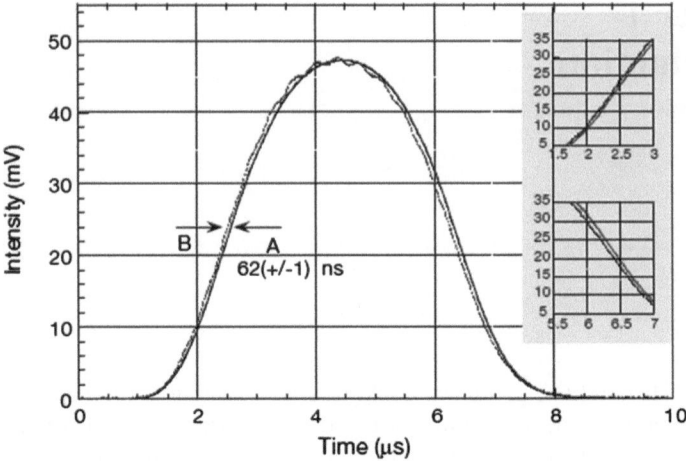

Fig. 1.1 Gaussian pulse propagation in the gain doublet medium, where the group velocity of the light pulse is negative. Reprinted by permission from Macmillan Publishers Ltd: Wang et al. [13]. Copyright 2013

the pulse was high enough to induce a nonlinear optical response and the pulse actually underwent severe distortion. Later, Garrett and McCumber [3] examined a weak Gaussian pulse propagating through the anomalous dispersion medium. With a long Gaussian-pulse and a short medium, they first reported theoretically that the pulse retained the shape while propagating with a negative group velocity. Large amount of theoretical interpretation and experimental reports enrich the discussions in the topic of fast light at the end of last century [4–12]. Beautiful experiments include the work of Wang et al. [13], which showed a tiny peak advance of 62 ns in the superluminal medium, with negligible pulse distortion and attenuation, as shown in Fig. 1.1. In their gain doublet medium, the Gaussian pulse maintained its original shape when passing through such a gain-assisted region. Later, Michael D. Stenner demonstrated the Gaussian propagation in the anomalous dispersion region of atomic ensembles [14], and verified that the information did not follow the advancement. On the other hand, Gehring et al. [15] observed backward pulse in the course of the forward energy flow in erbium-doped optical fiber. Other interpretation of superluminal velocity can still be found in recent works [16]. Actually, most scientists agree that the fast light observations do not violate Einstein special relativity theory, and rather, it is the result of classical EM field propagation theory.

To visualize the question, Fig. 1.2 displays the pulse propagation with a negative velocity. The pulse is emitted from the medium at the output surface before entering into the input surface. At the same time, a backward pulse travels to cancel the incoming pulse at the input surface. This behavior violates causality principle: the cause determines the results. Actually, Fig. 1.2 exaggerates the situation occurring in superluminal medium, where usually the propagation length

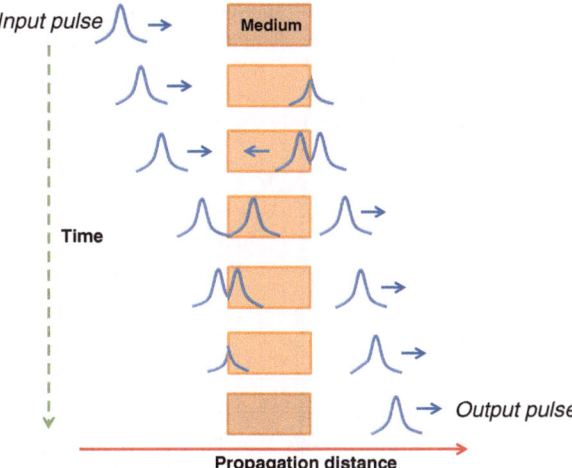

Fig. 1.2 Light pulse propagated with negative group velocity

of the medium is smaller than the pulse width. One can also describe it as pulse reshaping and energy redistribution [4] when the pulse is propagating in such kind of medium.

1.2 Single Photon Tunneling

Classical optical pulses are consisted of a large number of photons, following macroscopic electromagnetic wave propagation theory. While it is possible that a classical optical pulse travels with a negative group velocity, what about a photon in this pulse? Therefore, the problem is particularly interesting when the light pulse consists of a few photons, or even, a single photon. The speed of a classical light pulse represents the average speed of this group of photons, but does not guarantee every single one.

Figure 1.3 shows coincidence profile reported by Steinberg et al. [17] to demonstrate the seemingly "superluminal" propagation of a single photon with a Gaussian wave-packet. By scanning the trombone prism which varied the optical length for the single photons, they mapped out the single photon wave-packet. After inserting the barrier into the light path, they need to increase the optical length to obtain coincidence. In Fig. 1.3, they reported that a superluminal velocity $(1.7 \pm 0.2)c$ was observed. They interpreted the phenomenon as a result of "weak measurement", in which forbidden values could be obtained by both state preparation and postselection of low probability. This work does not rule out the probability for a photon traveling with speed faster than light. Quite a few comments have been published to review the photon tunneling effect, and discussed the true speed with which a photon travels [18, 19]. A number of experimental researcher believe that the causality principle must applies to single photons

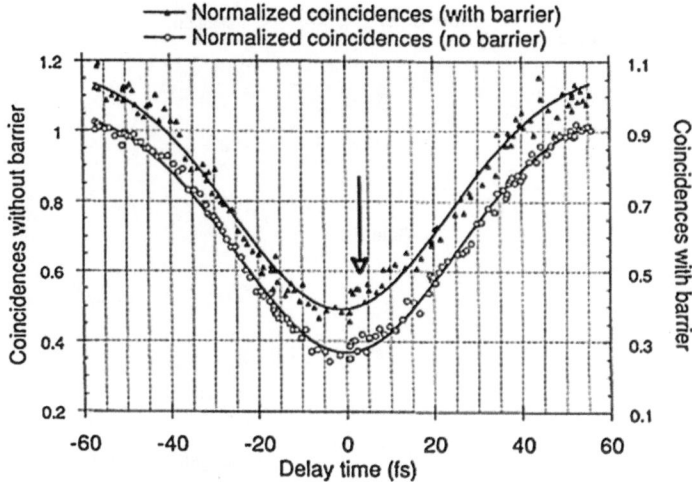

Fig. 1.3 Single photon wave packet advancement observed by Steinberg et al. Reprinted figure with permission from [17]. Copyright 2013 by the American Physical Society

[20, 21], but some argue that the group velocity maybe the same as the velocity of individual photons [18].

In Chaps. 5 and 6 in this book, we would like to answer this question, with experimental works concerning optical precursor. Before doing so, we would like to review the definitions for different velocities and their relations.

1.3 Phase Velocity, Group Velocity and Information Velocity

Phase velocity $v_p = \omega/k$ describes the propagation of a plane wave $\exp(-i\omega t + ikx)$ with a single optical frequency. But actually each optical pulse is consisted of a continuous frequency spectrum. In vacuum or air, the phase velocity is constant, equal to c. In dispersive medium, different spectral components in a wave train move with different phase velocities and thus the pulse distorts in some extent. However, when the distortion is slight enough and the wave train retains to be well-shaped, it is still meaningful to determine the group velocity. The group velocity in a dispersive medium is written as:

$$
\begin{aligned}
v_g &= \frac{d\omega}{dk}\Big|_0 \\
&= \frac{c}{n(\omega) + \omega(dn/d\omega)}
\end{aligned}
\tag{1.1}
$$

Usually, $dn/d\omega$ is positive, and thus v_g is smaller than the constant c. In an anomalous dispersion region, where the frequency of light coincide with the excitation resonance of the material, $dn/d\omega$ is a negative number. If $dn/d\omega$ is so negative that the term $\omega\frac{dn}{d\omega}$ is larger than $n(\omega)$, v_g takes a negative value.

Another remarkable type of medium is the material with electromagnetically-induced transparency (EIT) [22–24]. We can calculate v_g of the main signal propagating in an EIT system, from the linear susceptibility of EIT system in Refs. [25, 26]. Imagine the simplest case, where both the probe and coupling fields are on-resonance to the transitions. The linear susceptibility can be simplified as:

$$\chi(\omega) = \frac{OD}{k_{p0}L} \cdot \frac{4\gamma_{13}(\omega + i\gamma_{12})}{|\Omega_c|^2 - 4(\omega + i\gamma_{13})(\omega + i\gamma_{12})} \approx \frac{OD}{k_{p0}L} \frac{4(\omega + i\gamma_{12})\gamma_{13}}{\Omega_c^2 + 4\gamma_{12}\gamma_{13}} \tag{1.2}$$

And with the strong coupling field limit, the group velocity can be approximated as

$$v_g \approx \frac{2L}{OD} \frac{\Omega_c^2}{4\gamma_{13}} \tag{1.3}$$

In which $OD \equiv \alpha_0 L$ is the optical depth of medium, Ω_c^2 is the Rabi frequency of the coupling field and γ_{13} is the dephasing rate between the ground state and excited state. From Eq. (1.3), the main signal propagating in an EIT medium is much more slowed down due to an increase of optical depth. Also, the strength of the coupling field could modify the group velocity.

Extensive discussion concerning information has been made in various literatures, for example in Ref. [27]. Information is conveyed in the process that, the initial situation is the one with a certain number of equally possible outcomes, and a single outcome is selected in the final situation. In the binary unit system, each bit has a binary choice: 0 or 1. It is reasonable to define that one bit of information is sent if the amplitude or phase of the input pulse changes from "0" to "1", or "1" to "0". Similarly, if one detects the output pulse with a change of amplitude or phase, the information is received. For example, the amplitude of the pulse changes from 0 mV (defined as "0") to 5 mV (defined as "1"). The information is sent at the very front moment when the amplitude is still 0 mV, and is received when the amplitude increase to be 5 mV at the output detection side. If we define the time for the whole process is T, the information velocity is obtained from the effective propagation distance divided by T. Obviously, T includes the time for the changing of the input value. In principle, any pulse shape can be utilized to send information, as long as the amplitude or phase change can be identified. The most efficient way is to encode information in a non-analytic wave front, i.e., step-rising or falling edge, with which such a functional value change is completed at a single moment. To demonstrate the information velocity is different from the group velocity, Sommerfeld and Brillouin [28, 29] theoretically studied the propagation of a step pulse through a dispersive medium. They defined "signal" at the non-analytic wave front, and proved that the wave front of the step edge travelled at the fastest speed in a medium, the speed of light in vacuum c.

1.4 What is Precursor?

The concept of precursor was introduced by Sommerfeld and Brillouin in 1914 when they aimed to discuss the possibility to transmit signal in a dispersive medium with a speed faster than c. They pointed out that a signal should be effectively delivered with a non-analytical wave front, theoretically represented as a step pulse. In the dispersive medium which separates spectral constitutes, a small part of the incident step-pulse always propagates at the very front of the whole main pulse, with a velocity at or very close to the speed of light in vacuum. The small signal was called "forerunners", because they always exist at the very front of the whole bulk of transmitted pulse. This particular signal caught attentions from other physicists afterwards, and the name "precursor" was widely accepted later on. Specifically, two spectral poles contribute to the precursor field: extremely high-frequency ($\omega \rightarrow \infty$) Sommerfeld precursor and low-frequency ($\omega \rightarrow 0$) Brillouin precursor. Both of them are frequency components far away from the excitation resonance ω_0, and thus weakly interact with the ground-state atoms. The atoms are always transparent for these two branches of spectral components. In contrast, another catalogue of optical transients, free-induction decay (FID) [30–34], is referring to the frequency components at the vicinity of the atomic resonance. To describe in detail what happens inside the medium, we could view the atoms as a cloud of microscopic dipoles. Generally speaking, all of these optical transients are the consequences of light-matter interaction, in which a macroscopic polarization is created when the incident optical field polarizes the microscopic dipoles. We restrict the problem in a linear interaction effect, even though FID is not restricted to be a linear effect. Therefore we assume that the incident field is weak, and it can only stimulate a linear polarization response from the medium. The interaction process is described as Fig. 1.4. This linear polarization generates a radioactive field which interferes with the incident field. The modified incident field interacts with the macroscopic polarization again and the above process repeats until a steady state is obtained. Usually, the steady-state polarization takes a finite time to be established. Imagine an optical pulse with a sharp step front where the electromagnetic field suddenly turns on or turn off. If the cloud of dipoles is so intense that they severely absorb the resonant and near-resonant spectral components of the incident pulse, the main pulse could not transmit through the medium. However, the steep wave front survives because the dipoles do not have enough time to response to the sudden change of the incident field. Therefore the fronts transmit through the atoms as through the vacuum, i.e., propagating without loss and with the constant value c. This signal constitutes precursor. If the resonant and near-resonant spectral components are not absorbed

Fig. 1.4 Interaction between incident EM field and atoms. The figure is from Jeong [43]

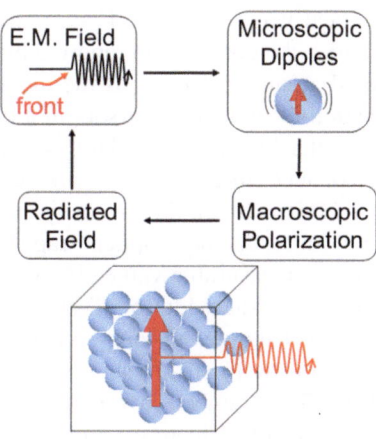

completely by the medium, their response to the changing excitation field produces the FID signal.

1.5   History of Precursor Research

In 1914, when Sommerfeld and Brillouin [28] started up the discussion of the various velocities involved in the propagation of a wave packet, they found out that the group velocity can be differed much from the signal velocity in the region where dispersion is anomalous. At the same time, in their analytical calculation using asymptotic evaluation of Fourier integrals, the precursor was predicted. Certain frequency constitutes of a finite pulse train will propagate through the medium with a velocity close to the speed of light in vacuum c. However, Sommerfeld and Brillouin made a mistake and predicted that the magnitude of precursor is of the order of 10^{-7} with extremely short coherent time of fs. Later on Oughstun and Sherman [35] identified the mistakes made in the early works, and pointed out that the precursor signals can be similar with the amplitude of main signal. Though the effect was well studied by various theoretical groups, the experimental observation of precursor remained a challenge due to the broad excitation resonance of the existing medium. The first direct experimental demonstration of Sommerfeld and Brillouin precursor was reported in 1969 by Pleshko and Palócz [36], in the microwave domain within a waveguide with artificially controllable stop band. The observation of optical SB-precursors is an even bigger challenge, since the optical frequency is extremely high and far beyond the technological limit of detector resolution. The first observation of SB-precursors in the optical regime is obtained with semiconductor GaAs, with the incident carrier-frequency set relatively close to an exitonic resonance [37]. Precursor behaviors have been reported in γ rays [38], microwaves [36], sound waves [39, 40], and optical waves [26, 37, 41–43] and single-photon correlations [44]. Among these works, the observation reported by Jeong et al. [43] and Wei et al. [26], introduced cold atomic ensemble as the medium. Jeong et al. reported the first direct observation of optical precursors in a cloud of cold potassium atomic gas (^{39}K). Since Doppler broadening is largely suppressed in the cold atom medium, the atomic resonance linewidth is extremely narrow, of the order of natural linewidth ($\sim$ MHz), which is inversely proportional to the precursor time scale. Thus, building a magneto-optical trap (MOT) is their key concept to increase time-scale up to a measurable value (16–32 ns). The optical precursors were then observable under the bandwidth of all electronics including the detectors. They demonstrated tens of nanosecond-long spikes with near 100 % transmission. The peaks at the front decay to steady state values due to the absorption of main signal expected from Beer-Lambert law. However, optical depth (OD) of potassium atoms was not enough to test mature regime (large OD) nor beating between two types of precursors. Consequently the precursor signal was not able to be separated from the main pulse part in their works. Motivated by their works in cold atom medium, the first observation of optical precursors through three-level EIT system was reported by Wei et al., Chen et al. [26, 45]. With the advantage of electromagnetically-induced transparency window and slow light effect, they observed the precursor and the main pulse presented together in a single measurement. The clear signature of optical precursor was presented in these measurements. The precursor is propagating in the medium with much less loss than the main pulse, and therefore it has potential application in teleportation and communication in dense material. Under-water communication, bio-imaging in tissues, for example, could utilize precursor as the signal carrier.

Optical precursor research in coherent cold atoms naturally triggers revisiting analogous quantum optical phenomena, such as 0π pulse [45, 46], optical nutation and free-induction decay. 0π pulse in area theorem [45] was compared to optical precursors theoretically [39, 47], but had been ignored by most of community until the comparison [48] with experimental demonstration of precursors in resonant regime [43] due to conventional ultra-fast characteristics. Dartmouth group compared all the formalism appeared in the relevant literatures together and show that the analytic expression of resonant precursors can be derived both by Maxwell's equations and optical Bloch equations [49].

Well-known optical transients, optical nutation and optical free-induction decay (FID), are optical analogues of spin resonance effect [50], and were first observed by Brewer and Shoemaker in 1972 [30, 31]. The equations describing such optical phenomena are formally equivalent to the well-studied spin-resonance Bloch equation. Optical FID was studied extensively in the early days, especially after the invention of the laser, and later was introduced in two-photon process [51–54]. All of these FID experiments in the past were weak and be indirectly detected by heterodyne means through the interference between radiation and the driving field. With the same problem similar with the observation of optical precursor, the broad linewidth of transition resonance and far-detuned driving field restrict the FID signal to be weak and fast. With modern technology of laser cooling and trapping, direct observation of FID signal was demonstrated within laser-cooled ^{85}Rb atoms [55]. In recent years, people work on unification of the concepts of optical precursors and coherent transients [56, 57]. Optical precursors and FIDs have generally been considered as two separate effects in the past decades. That's probably because the two phenomena occur under very different conditions of optical thickness and excitation strength. FIDs focus on the time-dependent phase coherence of the atomic states excited by the radiation. It is not restricted to linear dispersion theory and thus includes strong driving fields' cases. Rather, assumption of negligible propagation effects is taken while calculating FID fields. On the other hand, SB precursors are limited to weak excitation system, and thus frequency-domain linear transfer function play an important role in the theory. We will talk more about this in the next chapter.

References

1. Einstein, A.: On the electrodynamics of moving bodies. Ann. Phys. **17**, 891 (1905)
2. Basov, N.G., Ambartsumyan, R.V., Zuev, V.S., Kryukov, P.G., Letokhov, V.S.: Propagation velocity of an intense light pulse in a medium with inverse population. Sov. Phys. Doklady **10**, 1039 (1966)
3. Garrett, C.G.B., McCumber, D.E.: Propagation of a Gaussian pulse through an anomalous dispersion medium. Phys. Rev. A **1**, 305 (1970)
4. Crisp, M.D.: Concept of group velocity in resonant pulse propagation. Phys. Rev. A **4**, 2104 (1971)
5. Puri, A., Birman, J.L.: Pulse propagation in spatially dispersive media. Phys. Rev. A **27**, 1044 (1983)
6. Mache, B.: Vers une rehabilitation des vitesses de groupe negatives? Opt. Commun. **49**(5), 307–312 (1984)
7. Faxvog, F.R., Chow, C.N.Y., Bieber, T., Carruthers, J.A.: Measured pulse velocity greater than c in a Neon absorption cell. Appl. Phys. Lett. **17**, 192 (1970)
8. Chu, S., Wong, S.: Linear pulse propagation in an absorbing medium. Phys. Rev. Lett. **48**, 738–741 (1982)
9. Segard, B., Mache, B.: Observation of negative velocity pulse propagation. Phys. Lett. A **109**(5), 213–216 (1985)
10. Mitchell, M.W., Chiao, R.Y.: Causality and negative group delays in a simple bandpass amplifier. Am. J. Phys. **66**(1), 14–19 (1998)
11. Diener, G.: Superluminal group velocities and information transfer. Phys. Lett. A **223**, 327–331 (1996)
12. Hass, K., Busch, P.: Causality of superluminal barrier traversal. Phys. Lett. A **185**, 9–13 (1994)
13. Wang, L.J., Kuzmich, A., Dogariu, A.: Gain-assisted superluminal light propagation. Nature **406**, 277–279 (2000)
14. Stenner, M.D., Gauthier, D.J., Neifeld, M.A.: The speed of information in a "fast light" optical medium. Nature **425**, 695–698 (2003)
15. Gehring, G.M., Schweinsberg, A., Barsi, C., Kostinski, N., Boyd, R.W.: Observation of backward pulse propagation through a medium with a negative group velocity. Science **312**, 895–897 (2006)
16. Winful, H.G.: Nature of "superluminal" barrier tunneling. Phys. Rev. Lett. **90**, 023901 (2003)

17. Steinberg, A.M., Kwiat, P.G., Chiao, R.Y.: Measurement of the single-photon tunneling time. Phys. Rev. Lett. **71**, 708 (1993)
18. Marangos, J.: Faster than a speeding photon. Nature **406**, 243–244 (2000)
19. Büttiker, M., Washburn, S.: Ado about nothing much? Nature **422**, 271–272 (2003)
20. Chiao, R.Y.: Superluminal (but causal) propagation of wave packets in transparent media with inverted atomic populations. Phys. Rev. A **48**, R34 (1993)
21. Steinberg, A.M., Chiao, R.Y.: Dispersionless, highly superluminal propagation in a medium with a gain doublet. Phys. Rev. A **49**, 2071 (1994)
22. Harris, S.E.: Electromagnetically-induced transparency. Phys. Today **50**, 36–42 (1997)
23. Hau, L.V., Harris, S.E., Dutton, Z., Behroozi, C.H.: Light speed reduction to 17 metres per second in an ultracold atomic gas. Nature **397**, 594–598 (1999)
24. Fleischhauer, M., Imamoglu, A., Marangos, J.P.: Electromagnetically induced transparency: Optics in coherent media. Rev. Mod. Phys. **77**, 633–673 (2005)
25. Jeong, H., Du, S.: Two-way transparency in the light-matter interaction: Optical precursors with electromagnetically induced transparency. Phys. Rev. A **79**, 011802 (2009). (R)
26. Wei, D., Chen, J.F., Loy, M.M.T., Wong, G.K.L., Du, S.: Optical precursors with electromagnetically-induced transparency in cold atoms. Phys. Rev. Lett. **103**, 093602 (2009)
27. Brillouin, L.: Science and Information Theory. Academic press, New York (1956)
28. Sommerfeld, A.: Über die fortpflanzung des lichtes indispergierenden medien. Ann. Phys. **44**, 177 (1914)
29. Brillouin, L.: Über die fortpflanzung des lichtes in dispergierenden medien. ibid. **44**, 203 (1914)
30. Brewer, R.G., Shoemaker, R.L.: Photo echo and optical nutation in molecules. Phys. Rev. Lett. **27**, 631 (1971)
31. Brewer, R.G., Shoemaker, R.L.: Optical free induction decay. Phys. Rev. A **6**, 2001 (1972)
32. Foster, K.L., Stenholm, S., Brewer, R.G.: Interference pulses in optical free induction decay. Phys. Rev. A **10**, 2318 (1974)
33. Hopf, F.A., Shea, R.F., Scully, M.O.: Theory of optical free-induction decay and two-photon superradiance. Phys. Rev. A **7**, 2105 (1973)
34. Loy, M.M.T.: Observation of two-photon optical nutation and free-induction decay. Phys. Rev. Lett. **36**, 1454 (1976)
35. Oughstun, K.E., Sherman, G.C.: Propagation of electromagnetic pulses in a linear dispersive medium with absorption (the Lorentz medium). J. Opt. Soc. Am. B **5**(4), 817–849 (1988)
36. Pleshko, P., Palócz, I.: Experimental observation of Sommerfeld and Brillouin precursors in the microwave domain. Phys. Rev. Lett. **22**, 1201 (1969)
37. Aaviksoo, J., Kuhl, J., Ploog, K.: Observation of optical precursors at pulse propagation in GaAs. Phys. Rev. A **44**, R5353 (1991)
38. Lynch, F.J., Holland, R.E., Hamermesh, M.: Time dependence of resonantly filtered gamma rays from Fe57. Phys. Rev. **120**, 513 (1960)
39. Varoquaux, E., Williams, G.A., Avenel, O.: Pulse propagation in a resonant medium: Application to sound waves in superfluid ^{3}He-*B*. Phys. Rev. B **34**, 7617–7640 (1986)
40. Falcon, E., Laroche, C., Fauve, S.: Observation of Sommerfeld precursors on a fluid surface. Phys. Rev. Lett. **91**, 064502 (2003)
41. Sakai, M., Nakahara, R., Kawase, J., Kunugita, H., Ema, K.: Polariton pulse propagation at exciton resonance in CuCl: Polariton beat and optical precursor. Phys. Rev. B **66**, 033302 (2002)
42. Choi, S., Österberg, U.L.: Observation of optical precursors in water. Phys. Rev. Lett. **92**, 193903 (2004)
43. Jeong, H., Dawes, A.M.C., Gauthier, D.J.: Direct observation of optical precursors in a region of anomalous dispersion. Phys. Rev. Lett. **96**, 143901 (2006)
44. Du, S., Belthangady, C., Kolchin, P., Yin, G.Y., Harris, S.E.: Observation of optical precursors at the biphoton level. Opt. Lett. **33**, 2149 (2008)
45. Crisp, M.D.: Propagation of small-area pulses of coherent light through a resonant medium. Phys. Rev. A **1**, 1604–1611 (1970)

46. Rothenberg, J.E., Grischkowsky, D., Balant, A.C.: Observation of the formation of the 0π pulse. Phys. Rev. Letts. **53**(6), 552–555 (1984)
47. Avenel, O., Varoquaux, E., Williams, G.A.: Comment on observation of the formation of 0π pulse. Phys. Rev. Lett. **53**(21), 2058 (1984)
48. Jeong, H., Österberg, U.: Coherent transients: optical precursors and 0π pulses. J. Opt. Soc. Am. B **25**, B1–B5 (2008)
49. Lukofsky, D., Bessette, J., Jeong, H., Garmire, E., Österberg, U.: Can precursors improve the transmission of energy at optical frequencies. J. Mod. Opt. **56**(9), 1083–1090 (2009)
50. Bloch, F.: Nuclear Induction. Phys. Rev. **70**, 460 (1946)
51. Grischkowsky, D., Loy, M.M.T., Liao, P.F.: Adiabatic following model for two-photon transitions: Nonlinear mixing and pulse propagation. Phys. Rev. A **12**, 2514 (1975)
52. Liao, P.F., Bjorkholm, J.E., Gordon, J.P.: Observation of two-photon free-induction decay in atomic sodium vapor. Phys. Rev. Lett. **39**, 15 (1977)
53. Gold, D.G., Hahn, E.L.: Two-photon transient phenomena. Phys. Rev. A **16**, 324 (1977)
54. Lee, H.W.L., Wessel, J.E.: Observation of dressed-atom effects in three-level free-induction decay. Phys. Rev. Lett. **59**, 1416 (1987)
55. Toyoda, K., Takahashi, Y., Ishikawa, K., Yabuzaki, T.: Optical free-induction decay of laser-cooled 85Rb. Phys. Rev. A **56**, 1564 (1997)
56. Chen, J.F., Wang, S., Wei, D., Loy, M.M.T., Wong, G.K.L., Du, S.: Optical coherent transients in cold atoms: From free-induction decay to optical precursors. Phys. Rev. A **81**, 033844 (2010)
57. LeFew, W.R., Venakides, S., Gauthier, D.J.: Accurate description of optical precursors and their relation to weak-field coherent optical transients. Phys. Rev. A **79**, 063842 (2009)
58. Gauthier, D.J., Boyd, R.W.: Fast light, slow light and optical precursors: what does it all mean? Photonics Spectra **1**, 82–90 (2007)

Chapter 2
Theory of Optical Precursors

Abstract In this chapter, we discuss the theoretical framework of optical precursors based on the incident electromagnetic waves interacting with dielectric media. The simplest way to interpret the light-matter interaction is the medium optical response to the incident light characterized by dielectric constant $\varepsilon(\omega)$ as a function of incident light frequency ω. To build a theoretical model of optical precursors, first we derive macroscopic dielectric constant $\varepsilon(\omega)$ starting from the microscopic dipole moment $p(\omega)$. Based on Maxwell's equation and transfer function, the transmitted step-modulated electromagnetic field through dielectric media will be derived in general form of inverse Fourier transform of transmitted spectrum. The general expression of transmitted field can be solved numerically or analytically depending on the specific parameter regimes, such as Brillouin regime or resonant regime. The discussion extends from single Lorentz medium to electromagnetically-induced transparency medium, where the main signal transmits without loss.

2.1 Lorentz Medium and Transfer Function

Let's start with the conventional approach, in which we consider collection of dipole moments oscillating at the characteristic frequency ω_0 under externally shined electromagnetic field $E(x, t)$ in x direction as depicted by Fig. 2.1. This can be modeled as Lorentz oscillator driven by external force of E-field denoted as F,

$$
\begin{aligned}
F &= m\ddot{x} \\
&\Leftrightarrow -eE(x, t) - 2m\gamma\dot{x} - m\omega_0^2 x = m\ddot{x} \\
&\Leftrightarrow -eE(x, t) = m(\ddot{x} + 2\gamma\dot{x} + \omega_0^2 x)
\end{aligned}
\tag{2.1}
$$

where x denotes the displacement from the equilibrium position of the Lorentz oscillator. m denotes the mass of electron, and e denotes the electron charge. 2γ is

JF Chen et al., *Optical Precursors*, SpringerBriefs in Physics,
DOI: 10.1007/978-981-4451-94-9_2, © The Author(s) 2013

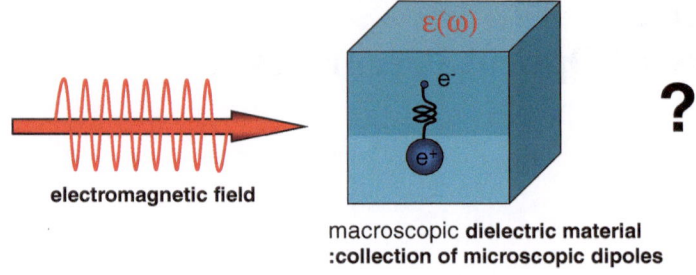

$$p(\omega) \Rightarrow P(\omega) \Rightarrow \chi(\omega) \Rightarrow \varepsilon(\omega)$$

Fig. 2.1 Light interaction with dielectric media

the damping constant, and ω_0 denotes resonant frequency. Equation (2.1) indicates the model follows damped harmonic oscillators governed by Hooke's law. Assume that the external electromagnetic field oscillates at ω. By Fourier transform, Eq. (2.1) is rewritten as displacement x.

$$-eE(x,\omega) = m(-\omega^2 - 2i\omega\gamma + \omega_0^2)x$$
$$x = \frac{eE(x,\omega)/m}{\omega^2 - \omega_0^2 + 2i\omega\gamma} \qquad (2.2)$$

Then microscopic dipole oscillator $p(\omega)$ and microscopic polarization $P(\omega)$ are obtained as a function of frequency,

$$p(\omega) = -ex = \frac{-e^2/m}{\omega^2 - \omega_0^2 + 2i\omega\gamma}E(x,\omega)$$
$$P(\omega) = Np(\omega) = \frac{-Ne^2/m}{\omega^2 - \omega_0^2 + 2i\omega\gamma}E(x,\omega) \qquad (2.3)$$

where N is the number of oscillators. Equation (2.3) implies how the macroscopic polarization is formed by the incident E-field. To relate $E(x,\omega)$ and $P(\omega)$, the linear susceptibility $\chi(\omega)$ can be introduced as

$$P(\omega) = \varepsilon_0\chi(\omega)E(x,\omega) \qquad (2.4)$$

where

$$\chi(\omega) = \frac{-Ne^2/m}{\omega^2 - \omega_0^2 + 2i\omega\gamma} \qquad (2.5)$$

Now the medium response, macroscopic polarization $P(\omega)$, is added to the original E-field to form a total displacement field $E(x,\omega)$.

$$D(x,\omega) = \varepsilon(\omega)E(x,\omega) = \varepsilon_0E(x,\omega) + P(\omega) = (1 + \chi(\omega))\varepsilon_0E(x,\omega) \qquad (2.6)$$

where $\varepsilon(\omega)$ is the dielectric function and the plasma frequency is ω_{pl}, and we have,

$$\varepsilon(\omega) = 1 + \chi(\omega) = 1 - \frac{\omega_{pl}^2}{\omega^2 - \omega_0^2 + 2i\omega\gamma} \qquad (2.7)$$

where $\omega_{pl} = \sqrt{Ne^2/\varepsilon_0 m}$. Dielectric function $\varepsilon(\omega)$ represents the medium response to an incident light. In general, the first term of $\varepsilon(\omega)$ is the background dielectric constant ε_0.

The dielectric function is the key to understand the physical interpretation of optical precursors. Plasma frequency ω_{pl} indicates the strength of absorption, one of the mechanisms in light-matter interaction. Full width half max 2γ implies the life time of the system, so that it affects the time scale of the transients or optical precursors. Finally, the atomic resonant frequency ω_0 with respect to the carrier frequency ω_p determines the field strength of the transients, i.e. optical precursors.

In the next section, the dielectric function of a Lorentz medium plays a role in the transfer function, $T(z, \omega) = e^{ik(\omega)z}$, as shown in Fig. 2.2, to evaluate emerging field $E(z, t)$ out of the dielectric material.

To deal with the optical field propagation, we first unify the notation of the optical field throughout the whole book. The real electric field in plane wave is expressed as,

$$\vec{\mathbf{E}}(z, t) = \frac{1}{2} \mathrm{Re} \left\{ E_{evp}(z, t) \vec{n} e^{i(kz - \omega t)} \right\} \qquad (2.8)$$

where $\vec{n}$ is the polarization unit vector, and $E_{evp}(z, t)$ is the complex envelope. Fourier transform of the complex envelope gives the spectrum of the optical field,

$$\tilde{\mathbf{E}}(\omega, z) = \frac{1}{\sqrt{2\pi}} \int_{-\infty}^{\infty} \mathbf{E}(t, z) e^{i\omega t} dt \qquad (2.9)$$

Now let's consider an incident field $\vec{\mathbf{E}}(0, t)$ and consequent emerging field $\vec{\mathbf{E}}(z, t)$ out of dielectric medium. The medium is characterized via the transfer function $T(z, \omega) = e^{ik(\omega)z}$, which is derived from Maxwell's equation,

Fig. 2.2 Schematic diagram describing transfer function $\varepsilon(\omega)$

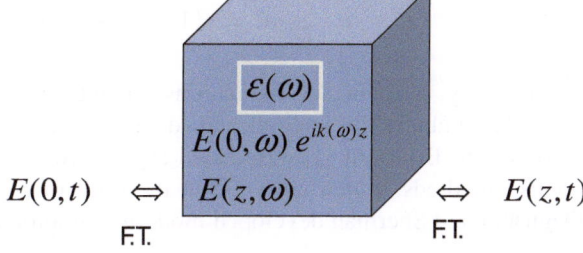

$E(0,t) \quad \Leftrightarrow \quad E(z,\omega) \qquad \Leftrightarrow \quad E(z,t)$

F.T.   F.T.

$$\vec{\nabla}^2 \vec{E}(z,t) - \frac{1}{c^2}\frac{\partial^2 \vec{E}(z,t)}{\partial t^2} = \frac{1}{\varepsilon_0 c^2}\frac{\partial^2 \vec{P}(z,t)}{\partial t^2} \tag{2.10}$$

According to Eq. (2.10), the polarization $\vec{P}(z,t)$ acts as a source of total electric field. Let's remind of the fact that the total field consists of the incident electric field $\vec{E}(z,t)$ and the modified one by the interaction with the medium polarization. We assume that a polarized plane-wave field propagation vector along the z-direction and isotropic medium and for convenience we denote the electric field as $E(z,t)$. The propagation of light is described by the scalar 1-dimensional wave equation, which is given as follows:

$$\frac{\partial^2 E(z,t)}{\partial z^2} - \frac{1}{c^2}\frac{\partial^2 E(z,t)}{\partial t^2} = \frac{1}{\varepsilon_0 c^2}\frac{\partial^2 P(z,t)}{\partial t^2} \tag{2.11}$$

Here, by performing Fourier transform on Eq. (2.11), it is easy to obtain explicit form of the medium response as a function of frequency and the spectrum of the electric field $\tilde{E}(0,\omega)$.

$$\frac{\partial^2 E(z,\omega)}{\partial z^2} + \frac{\omega^2}{c^2}(1+\chi(\omega))E(z,\omega) = 0$$

$$\Rightarrow \frac{\partial^2 E(z,\omega)}{\partial z^2} + k^2(\omega)E(z,\omega) = 0 \tag{2.12}$$

where $k(\omega) \equiv \omega\sqrt{\mu\varepsilon(\omega)}$. Then we express propagated field as a function of frequency.

$$E(z,\omega) = E(0,\omega)\,e^{ik(\omega)z} \equiv E(0,\omega)\,T(z,\omega) \tag{2.13}$$

By performing inverse Fourier transform, we could evaluate the transmitted field through medium as,

$$E(z,t) = \frac{1}{\sqrt{2\pi}}\int_{-\infty}^{\infty} E(z,\omega)\,e^{-i\omega t}d\omega = \frac{1}{\sqrt{2\pi}}\int_{-\infty}^{\infty} E(0,\omega)\,e^{i(k(\omega)-\omega t)}d\omega$$

$$= \frac{1}{\sqrt{2\pi}}\int_{-\infty}^{\infty} E(0,\omega)\,e^{\phi(\omega,\theta)}d\omega \tag{2.14}$$

where $\phi(\omega,\theta) = i\omega\frac{z}{c}(n(\omega)-\theta)$, $\theta \equiv \frac{ct}{z}$

$$n(\omega) = \sqrt{1 - \frac{\omega_{pl}^2}{\omega^2 - \omega_0^2 + 2i\omega\gamma}} \tag{2.15}$$

Generally, without approximations, analytic solution of Eq. (2.14) does not exist. To evaluate the integral and identify optical precursor components, Sommerfeld and Brillouin [1] introduced asymptotic theory associated with "saddle-points" methods, which are valid in the limit of $z \to \infty$. After several decades, Oughstun and Sherman developed modern asymptotic theory of optical precursors

[2]. The conventional asymptotic theory of optical precursors by SB and OS usually consider highly dissipative ($\gamma \sim 0.1\,\omega_0$) and off-resonance condition ($\omega_p \neq \omega_0$), which result in precursor transmission with small intensity and a femtosecond time scale. The modern asymptotic analysis introduced by OS usually shows complicated expression of precursors field which can be numerically evaluated.

Recently, on the other hand, recent experimental works report optical precursors [3, 4] in "resonant regime" [5], indicating the characteristics of on-resonant condition as well as narrow linewidth. The existence of optical precursors has been verified in the resonant regime, where one can obtain analytic expression of optical precursors [6]. To prove the existence of optical precursors within the boundary of SB and OS, LeFew simplify the OS's modern asymptotic theory as we will discuss later [6].

2.2 Classical Theory of Optical Precursors: Asymptotic Method

In the conventional theory of optical precursors, the input field is taken as a step-modulated sinusoidal electric field of the form as illustrated in Fig. 2.3.

$$E(0,t) = E_0\Theta(t)e^{-i\omega_p t} \tag{2.16}$$

where $\Theta(t)$ is the Heaviside unit step function. By performing Fourier Transform of the input pulse Eq. (2.16), the input spectrum $E(0,\omega)$ is

$$E(0,t) = E_0\Theta(t)e^{-i\omega_p t} \Leftrightarrow \tilde{E}(0,\omega) = \frac{iE_0}{\sqrt{2\pi}(\omega - \omega_p)} \tag{2.17}$$

The step-modulated electric field is the starting point of the theories. Optical precursor theory for other input pulses, such as a hyperbolic tangent-modulated pulse [2], square pulse [7], Gaussian pulse [8], have been discussed. In this section, we only consider a step-modulated or square-modulated input pulse.

Fig. 2.3 Step modulated optical field interacts with a Lorentz dielectric medium

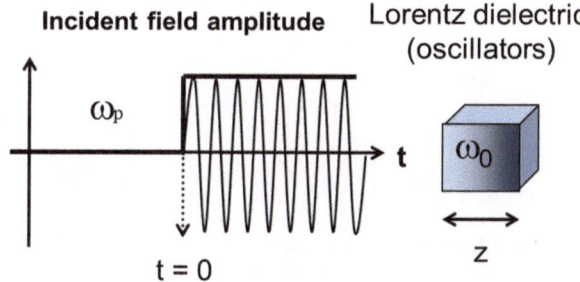

The modern asymptotic theory of OS is very complicated, so that it is difficult to obtain a physical meaning of the problem. The numerical evaluation over broad range of parameter spaces as well as understanding their characteristics will provide us insight into the propagation of transient pulses. Conventional method to handle the general type of integral Eq. (2.14) is asymptotic analysis based on the saddle-point method [9] which is valid in the limit when a distance into the medium $z \gg \alpha_0^{-1}$. The inverse of absorption coefficient α_0^{-1} indicates the distance over which the incident electromagnetic field intensity decreases by 1/e of its initial value.

To understand some of the concepts underlying the saddle-points method, let's start with presenting the "stationary phase" approximation in the following integral,

$$I = \int f(\omega)\, e^{iq(\omega)} d\omega \qquad (2.18)$$

where $f(\omega)$ is a slowly varying function in terms of ω. The phase term $q(\omega)$ is large enough, causing rapid scillation of the integrand $s^{iq(\omega)}$. In Fig. 2.4a, integration of the fast-oscillating field over the entire range of frequency ω is averaged out and thus result in zero-value except for slowly varying $q(\omega)$ near the stationary point, ω_{sp}, as illustrated in Fig. 2.4b. Note that the subscript 'sp' denotes the "stationary point", but will be used as "saddle point". To obtain the stationary points, the first derivative of the phase $\partial_\omega q(\omega)|_{\omega_{sp}} = 0$ is required. Phase has an extreme value at these stationary points $q(\omega_{sp})$, so that it is called "stationary phase". A non-zero contribution to the integral can be obtained by the integration along the stationary point (ω_{sp}). The "steepest-decent" method is a subset of the saddle point method for the case when the phase $q(\omega)$ is a real number. The stationary-phase approximation only considers the first leading-order term of the "steepest-decent" method.

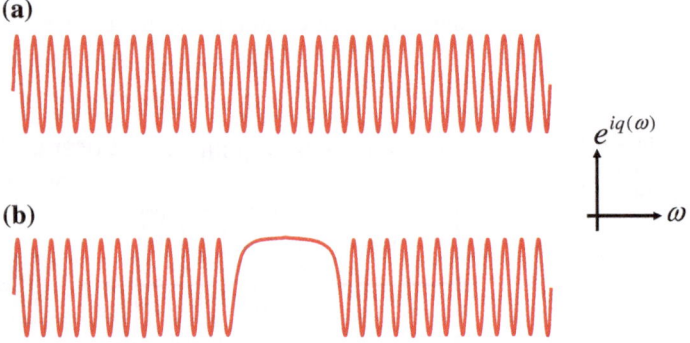

Fig. 2.4 Illustration of stationary phase. **a** Rapid oscillation without a stationary point, and **b** with a stationary point. The figure is from Jeong [9]

The above description provides us precise understanding of the concept of saddle-points in non-zero transmitted field components. Equation (2.14) is equivalent to the Eq. (2.18) when we let $f(\omega) \equiv \tilde{E}(0, \omega) = \frac{iE_0}{\sqrt{2\pi}(\omega - \omega_p)}$, and $q(\omega) \equiv z\varphi(\omega, \theta)/c$. Therefore, the saddle points, ω_{sp}, are obtained by the first derivative with respect to ω

$$\phi'(\omega_{sp}(z, t)) \equiv \partial_\omega \phi(\omega, \theta)|_{\omega_{sp}} = 0 \qquad (2.19)$$

The integral has a non-zero value when it is evaluated near the extreme value of the phase.

With the saddle-points ω_{sp} obtained by Eq. (2.19), the corresponding phases $\varphi(\omega_{sp}, \theta)$, and their second derivatives $\varphi''(\omega_{SP}(z, t)) \equiv \partial_\omega^2 \varphi(\omega, \theta)|_{\omega_{sp}}$, the phase can be expressed as Taylor expansion

$$\phi(\omega, \theta) \approx \phi(\omega_{sp}, \theta) + \phi'(\omega_{sp}(z, t))(\omega - \omega_{sp}) + \frac{1}{2!}\phi''(\omega_{sp}(z, t))(\omega - \omega_{sp})^2 \qquad (2.20)$$

and

$$\int_{-\infty}^{\infty} e^{\phi(\omega, \theta)} \, d\omega \approx \int_{-\infty}^{\infty} e^{\phi(\omega_{sp}, \theta) + \frac{1}{2!}\phi''(\omega_{sp}(z, t))(\omega - \omega_{sp})^2} \, d\omega = e^{\phi(\omega_{sp}, \theta)} \frac{\sqrt{2\pi}}{\sqrt{-|\phi''(\omega_{sp}(z, t))|}} \qquad (2.21)$$

by recalling $\int_{-\infty}^{\infty} e^{-\frac{\omega^2}{2\sigma^2}} d\omega = \sqrt{2\pi}\sigma$, where $\sigma \equiv 1/\sqrt{-|\phi''(\omega_{sp}(z, t))|}$. Therefore, Eq. (2.14) is given by

$$E_{\omega_{SP}}(z, t) = \frac{iE_0}{\sqrt{2\pi}(\omega_{SP} - \omega_c)} \frac{e^{\phi(\omega_{SP}, t) + i\psi}}{\sqrt{|\phi''(\omega_{SP}(z, t))|}} \qquad (2.22)$$

where ψ is the angle of steepest decent.

For the case of a single-resonance Lorentz medium, there are two types of saddle points as we will see in this chapter. The two saddle points are related to the two transient parts of the emerging transmitted field. One of two saddle points has high-frequency components associated with Sommerfeld precursor $E_s(z, t)$, and the other is low frequency or DC component associated with Brillouin precursors $E_B(z, t)$.

As one might notice, pole contribution to the integral becomes dominant and non-zero value can be obtained at singular point $\omega = \omega_p$. This pole contribution is associated with the steady-state part of the emerging field, main signal $E_C(z, t)$ as we will see later. To evaluate transient transmitted field (SB precursors), Oughstun and Sherman [2] have developed Brillouin's asymptotic analysis by keeping a higher order term in the saddle-point equation. OS also have used modern mathematical methods "Olver-type path" [10] to search for a convenient path of integral.

Since the original work by OS [2] have demonstrated saddle points and each part of transmitted field with complicated mathematical expression, we would like to deal with the simplest expression of the phase and consequent derivatives most recently suggested by LeFew [11]. Let's again consider the phase of integrand in Eq. (2.14).

$$\varphi(\omega) = \frac{z}{c} i\omega[n(\omega) - \theta] = \frac{z}{c} i\omega \left[\sqrt{\frac{\omega^2 - (\omega_0^2 + \omega_{pl}^2) + 2i\omega\gamma}{\omega^2 - \omega_0^2 + 2i\omega\gamma}} - \theta \right] \qquad (2.23)$$

Instead of using the conventional form, the complex frequency is set as $\eta \equiv i(\omega + i\gamma)$, which is shifted by $-\gamma$ from the imaginary axis of complex frequency ω and rotated by $90°$ [11]. Then, Eq. (2.23) can be simplified as

$$\varphi(\omega) = \frac{z}{c}(\eta + \gamma) \left[\frac{R_2}{R_1} - \theta \right] \qquad (2.24)$$

where $R_1 \equiv \sqrt{(\eta^2 + \omega_0^2 - \gamma^2)/\omega_0^2}$, and $R_2 \equiv \sqrt{(\eta^2 + \omega_0^2 + \omega_{pl}^2 - \gamma^2)/\omega_0^2}$. The saddle-point equation is given by

$$\partial_\omega \phi(\omega, \theta)|_{\omega_{sp}} = R_1^3 R_2 \theta - R_1^2 R_2^2 + (R_2^2 - R_1^2)\eta(\eta + \gamma)|_{\eta_{sp}} = 0 \qquad (2.25)$$

With four saddle-points, $\eta_{sp}^\pm (\eta_s^\pm$ and $\eta_B^\pm)$ or $\omega_{sp}^\pm (\omega_s^\pm$ and $\omega_B^\pm)$, obtained from Eq. (2.25), one can evaluate $E_S(z,t)$ and $E_B(z,t)$ numerically from Eq. (2.22) as

$$E_S(z,t) \approx \sum_{\omega_S^\pm} \frac{\mp E_0 e^{\varphi(\omega_S^\pm(z,t),t) - \frac{i}{2}Arg[\varphi''(\omega_S^\pm(z,t))]}}{\sqrt{2\pi}(\omega_S^\pm - \omega_p)\sqrt{|\phi''(\omega_S^\pm(z,t))|}} \qquad (2.26)$$

$$E_B(z,t) \approx \sum_{\omega_B^\pm} \frac{\mp E_0 e^{\phi(\omega_B^\pm(z,t),t) - \frac{i}{2}Arg[\phi''(\omega_B^\pm(z,t))]}}{\sqrt{2\pi}(\omega_B^\pm - \omega_p)\sqrt{|\phi''(\omega_B^\pm(z,t))|}} \qquad (2.27)$$

Besides the saddle-points, another non-zero contribution to the integral Eq. (2.14) arises from $1/(\omega - \omega_p)$ at singular point $\omega = \omega_p$. The pole contribution to the integral is related to the steady-state response of the medium to the incident field, which is known as the main signal.

$$E_C(z,t) = 2\pi i \, Res[\omega = \omega_p] \qquad (2.28)$$

With the numerical frame work of asymptotic theory: Eqs. (2.26–2.28), the total transmitted field intensity can be obtained as in Fig. 2.5. The left colum shows the results of asymptotic theory in Eqs. (2.26–2.28), which are compared to analytic results based on (2.48–2.49) [right column]. Although the recent asymptotic analysis [11] has significantly reduced the numerical errors for resonant regime, one cannot avoid intrinsic difference between the theory and the experimental data right at the front edge [12].

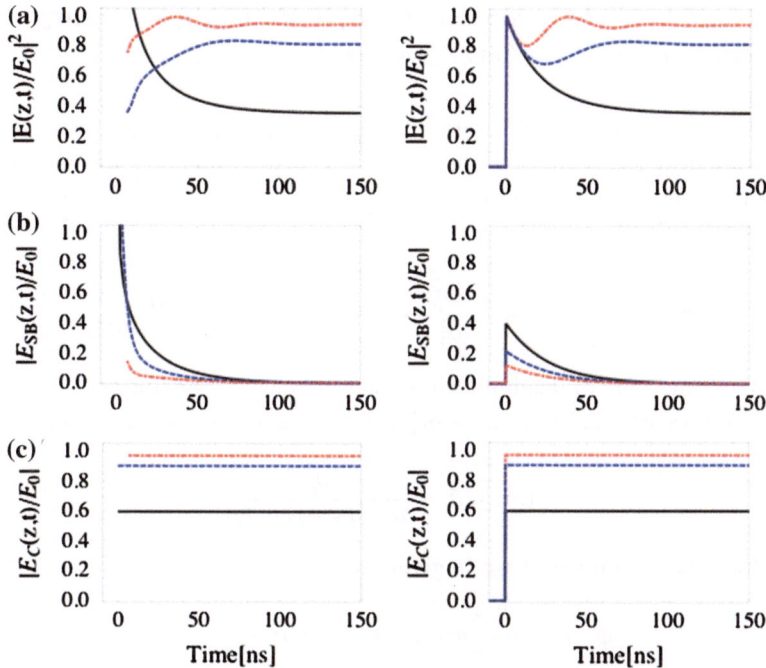

Fig. 2.5 The absolute value of total transient field envelope for the asymptotic theory [*left column*, Eqs. (2.26–2.28)] and for analytic expression [*right column*, Eqs. (2.48–2.49)]. **a** The total transient transmission. **b** The amplitude of precursors, and **c** main signal for $\Delta \sim 4\gamma$ (denoted by *red dash-dot line*), $\Delta \sim 2\gamma$ (denoted by *blue dashed line*), $\Delta \sim 0$ (denoted by *black solid line*)

2.3 Optical Precursor Theory for Resonant Medium

The main assumptions of the resonant regime are small plasma frequency ($\omega_{pl} \ll \sqrt{8\omega_0\gamma}$), narrow material resonance ($\gamma \ll \omega_0$), nearly resonance with material oscillators ($\omega_p \sim \omega_0$), and slowly varying approximation (SVA) [13–15]. Under these assumptions, it is possible to evaluate analytic solution describing the propagation of the step-modulated field through the single –Lorentz dielectrics [14–16]. The total emerging field is given by

$$E(z, t) = E_{SB}(z, t) + E_C(z, t) \tag{2.29}$$

where the total transient response is $E_{SB}(z, t)$, which should be equivalent to the sum of two precursors $E_S(z, t) + E_B(z, t)$.

2.3.1 Analytic Expression for a Single-Resonance Lorentz Dielectrics: Two-Level System

In this section, let's consider weakly dispersive resonant medium for which analytic solution of Eq. (2.14) is achievable. The first assumption we take is small plasma frequency condition, $\omega_{pl} \ll \sqrt{8\omega_0\gamma}$, to eliminate square root by the Taylor expansion,

$$n(\omega) = \sqrt{1 - \frac{\omega_{pl}^2}{\omega^2 - \omega_0^2 + 2i\omega\gamma}} \approx 1 - \frac{1}{2}\frac{\omega_{pl}^2}{\omega^2 - \omega_0^2 + 2i\omega\gamma} \qquad (2.30)$$

By the condition of medium's resonant frequency, we further simplify the denominator of Eq. (2.30) as $\omega^2 - \omega_0^2 + 2i\omega\gamma \approx 2\omega(\omega - \omega_0 + i\gamma)$, and hence

$$n(\omega) \approx 1 - \frac{\omega_{pl}^2}{4\omega(\omega - \omega_0 + i\gamma)} \qquad (2.31)$$

Therefore, the simplified phase is given as

$$\phi(\omega) = i\omega[\frac{z}{c}n(\omega) - t] \approx -i\omega\tau - \frac{ip}{\Delta_0 + i\gamma} \qquad (2.32)$$

where $p \equiv \alpha_0 z\gamma/2$, retarded time $\tau \equiv t - z/c$, detuning from medium resonance $\Delta_0 \equiv \omega - \omega_0$, and absorption coefficient $\alpha_0 \equiv \omega_{pl}^2/2\gamma c$. Based on the simplified phase Eq. (2.32), there are two ways to obtain analytic expression. One is contour integral which is used when we deal with off-resonant expression, but lose information about the saddle points. The other is method of steepest decent [17] and saddle-point method for the on-resonance case.

Method 1: Contour integral by Cauchy integral formula [16].

The first approach to solve Eq. (2.14) associated with simplified phase Eq. (2.32) is contour integral based on Cauchy integral formula [16] as illustrated in Fig. 2.6.

The original integral in Eq. (2.14) is performed along the real axis of the complex frequency plane in Fig. 2.6. By Cauchy theorem [17] the integral is given as

$$\begin{aligned}
E(z,t) &= \frac{E_0}{2\pi i}\int_{-\infty}^{\infty}\frac{e^{-i\omega\tau}}{\omega - \omega_p}e^{\frac{-i\alpha_0 z\gamma/2}{\omega - \omega_0 + i\gamma}}d\omega \\
&= \frac{E_0}{2\pi i}\oint_{\omega_p} Gd\omega + \frac{E_0}{2\pi i}\oint_{\omega_0 - i\gamma} Gd\omega \qquad (2.33) \\
&= E_c(z,t) + E_{SB}(z,t)
\end{aligned}$$

Fig. 2.6 Schematics of
contour integral

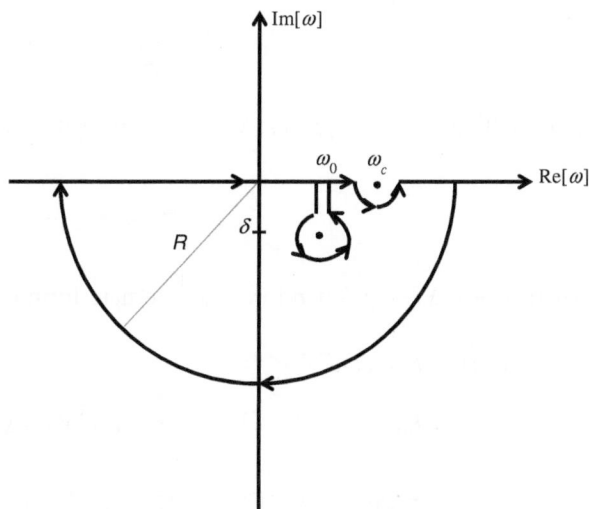

where the integrand is

$$G \equiv \frac{e^{i\omega\tau}}{\omega - \omega_p} e^{\frac{-i\alpha_0 z\gamma/2}{\omega - \omega_0 + i\gamma}} \tag{2.34}$$

The contour integral is divided into two associated with two singular points. First part is singular point at carrier frequency and it contribute to first term of the integral

$$E_C(z,t) = \frac{E_0}{2\pi i} \oint_{\omega_p} \frac{e^{-i\omega\tau}}{\omega - \omega_p} e^{\frac{-i\alpha_0 z\gamma/2}{\omega - \omega_0 + i\gamma}} d\omega \tag{2.35}$$

By setting the complex variable $v \equiv \omega - \omega_p$,

$$E_C(z,t) = \frac{E_0}{2\pi i} \oint_{v=0} \frac{dv}{v} e^{-i(v+\omega_p)\tau} e^{\frac{-i\alpha_0 z\gamma/2}{v + i\gamma}} \tag{2.36}$$

A solution to above expression can be evaluated by the residue theorem,

$$\oint_C f(v) dv = 2\pi i \sum (\text{enclosed residues}) \tag{2.37}$$

Thus, the solution has an analytic form indicating exponential decay of the envelope of the transmitted main signal.

$$E_C(z,t) = E_0 \Theta(\tau) e^{-i\omega_p \tau} e^{\frac{\alpha_0 z\gamma/2}{i\Delta_0 - \gamma}} \tag{2.38}$$

There is different way to express Eq. (2.38) using "generating Bessel function",

$$e^{\frac{x}{2}(u-\frac{1}{u})} = \sum_{m=-\infty}^{\infty} u^m J_m(x) \tag{2.39}$$

By letting $u = iy\sqrt{\tau/p}$, and $x = 2\sqrt{p\tau}$, we get different expression of Eq. (2.39) as

$$e^{i(\tau y - \frac{p}{y})} = \sum_{m=-\infty}^{\infty} i^m (\frac{\tau}{p})^{\frac{m}{2}} y^m J_m(2\sqrt{p\tau}) \tag{2.40}$$

where $y \equiv -\Delta_0 - i\gamma$. Therefore, the alternate form of Eq. (2.40) is given as

$$E_C(z,t) = E_0 \Theta(\tau) e^{\frac{\alpha_0 z\gamma/2}{i\Delta_0 - \gamma}} e^{-i\omega_p \tau}$$

$$= E_0 \Theta(\tau) e^{-i\omega_p \tau + (i\Delta_0 - \gamma)\tau} \sum_{m=-\infty}^{\infty} i^m (\frac{\tau}{p})^{\frac{m}{2}} (-\Delta_0 - i\gamma)^m J_m(2\sqrt{p\tau}) \tag{2.41}$$

$$= E_0 \Theta(\tau) e^{-i\omega_p \tau + (i\Delta_0 - \gamma)\tau} \sum_{n=-\infty}^{\infty} (\frac{p}{i\Delta_0 - \gamma})^n (p\tau)^{-n/2} J_n(2\sqrt{p\tau})$$

Now, let's look at the second term of Eq. (2.33), which is the contour integral around singular point of the exponent, $\omega_0 - i\gamma$, by setting $z \equiv \omega - \omega_p + i\gamma$.

$$E_{SB}(z,t) = \frac{E_0}{2\pi i} e^{-i(\omega_0 - i\gamma)\tau} \oint_{z=0} dz \frac{e^{-iz\tau - i\alpha_0 z\gamma/2z}}{z + \omega_0 - \omega_p - i\gamma} \tag{2.42}$$

By rewriting the denominator as

$$\frac{1}{z-a} = -\frac{1}{a}\sum_{n=0}^{\infty} (\frac{z}{a}) = -\sum_{n=0}^{\infty} \frac{z^n}{a^{n+1}} \tag{2.43}$$

where $a \equiv \omega_p - \omega_0 + i\gamma$, and

$$e^{-iz\tau - ip/z} = \sum_{m=-\infty}^{m=\infty} (-iz)^m (\frac{\tau}{p})^{m/2} J_m(2\sqrt{p\tau}) \tag{2.44}$$

where $p \equiv \alpha_0 z\delta/2$, Then we rewrite Eq. (2.33) as

$$E_{SB}(z,t) = -\frac{E_0}{2\pi i} e^{(-i\omega_0 - \gamma)\tau} \oint_{z=0} dz \sum_{m=-\infty}^{\infty} \sum_{n=0}^{\infty} \frac{z^{m+n} i^m (\tau/p)^{m/2} (-1)^m J_m(2\sqrt{p\tau})}{(\omega_p - \omega_0 - i\gamma)^{n+1}} \tag{2.45}$$

By considering the residue theorem

$$\int_{z=0} dz z^{m+n} = 2\pi i \delta_{m,(n+1)} \tag{2.46}$$

and $J_{-n}(x) = (-1)^n J_n(x)$,

$$E_{SB}(z,t) = -E_0 \Theta(\tau) e^{-i\omega_p \tau + (i\Delta - \gamma)\tau} \sum_{n=1}^{\infty} \left(\frac{p}{i\Delta_0 - \gamma} \right) (p\tau)^{-n/2} J_n(2\sqrt{p\tau}) \qquad (2.47)$$

for $\sqrt{p/(\tau(\Delta_0^2 + \gamma^2))} < 1$. Exponential decay $e^{-\gamma\tau}$ and the Bessel function $J_{-n}(2\sqrt{p\tau})$ in Eq. (2.47) affect the transient time scale, and it indicates that the second term of the contour integral is the transient part of total transmission.

Therefore, for $\sqrt{p/(\tau(\Delta_0^2 + \gamma^2))} < 1$, the total transmitted field is evaluated from the sum of Eqs. (2.41) and (2.47)

$$E_{SB}(z,t) = E_0 \Theta(\tau) e^{-i\omega_p \tau} \left(e^{\frac{p}{i\Delta_0 - \gamma}} - e^{(i\Delta - \gamma)\tau} \sum_{n=1}^{\infty} \left(\frac{p}{i\Delta_0 - \gamma} \right) (p\tau)^{-n/2} J_n(2\sqrt{p\tau}) \right)$$
$$(2.48)$$

The alternate form of the total transmission is derived by considering

$$\sum_{n=1}^{\infty} S_n = \sum_{n=-\infty}^{\infty} S_n - \sum_{n=-\infty}^{0} S_n, \qquad \text{and} \qquad \sum_{n=-\infty}^{\infty} S_n = \sum_{n=0}^{\infty} S_{-n} \qquad \text{where}$$

$S_n \equiv \left(\frac{p}{i\Delta_0 - \gamma} \right)^n (p\tau)^{-n/2} J_n(2\sqrt{p\tau})$. So in the other region where $\sqrt{p/(\tau(\Delta_0^2 + \gamma^2))} > 1$, we have,

$$E_{SB}(z,t) = E_0 \Theta(\tau) e^{-i\omega_p \tau + (i\Delta_0 - \gamma)\tau} \left(-\sum_{n=-\infty}^{\infty} S_n + \sum_{n=-\infty}^{\infty} S_n - \sum_{n=-\infty}^{0} S_n \right)$$
$$= E_0 \Theta(\tau) e^{-i\omega_p \tau + (i\Delta_0 - \gamma)\tau} \sum_{n=0}^{\infty} \left(\frac{-i\Delta_0 + \gamma}{p} \right)^n (p\tau)^{n/2} J_n(2\sqrt{p\tau}) \qquad (2.49)$$

The above expression for total transmitted field is useful if one would like to see the effect of detuning within near resonance regime, i.e., $\Delta_0 \approx \gamma$. However, $E_{SB}(z,t)$ cannot be separated to $E_S(z,t)$ and $E_B(z,t)$. To obtain analytic form of each precursor part, one requires restricted on-resonant condition of $\omega_p = \omega_0$ as discussed in method 2.

Method 2: Saddle-point approximation.

Let's redefine the phase $\phi(\omega, t)$ as $E(t) = \frac{1}{\sqrt{2\pi}} \int_{-\infty}^{\infty} E(0, \omega) e^{i\phi(\omega,\theta)} d\omega$, and the first and second derivatives of the phase are

$$\phi'(\omega_{sp}^{\pm}, t) \approx -\tau + \frac{p}{(\Delta_{sp} + i\gamma)^2}$$
$$\phi''(\omega_{sp}^{\pm}, t) \approx -\frac{2p}{(\Delta_{sp} + i\gamma)^3} \qquad (2.50)$$

From above equation, saddle-point can easily obtained from $\partial_\omega \phi(\omega, t)|_{\omega_{sp}} = 0$

$$\Delta_{sp}^{\pm} = \pm\sqrt{\frac{\alpha_0 z\gamma}{2\tau}} - i\gamma \tag{2.51}$$

where, $\Delta_{sp} \equiv \omega_{sp} - \omega_0$. Here, let's define $\xi(t) \equiv \sqrt{\frac{\alpha_0 z\gamma}{2\tau}} = \sqrt{\frac{\alpha_0 z\gamma}{2(t-L/c)}}$, then the de-tuned saddle-points is

$$\Delta_{sp}^{\pm} = \pm\xi(t) - i\gamma \tag{2.52}$$

and the second derivatives of the phase are written as

$$\phi''(\omega_{sp}^{\pm}) = -\frac{\alpha_0 z\gamma}{(\pm\xi(t))^3} \tag{2.53}$$

Now let's consider the general form of the transmitted field. The transmitted field is mainly attributed to the saddle point ω_{sp} contribution.

$$
\begin{aligned}
E(z,t) &\approx \frac{1}{\sqrt{2\pi}}\int_{-\infty}^{\infty} E(0,\omega_{sp})\, e^{i[\phi(\omega_{sp})+\frac{1}{2}\phi''(\omega_{sp})(\omega-\omega_{sp})^2]}d\omega \\
&= \frac{E(0,\omega_{sp})}{\sqrt{2\pi}}e^{i\phi(\omega_{sp},t)}\int_{-\infty}^{\infty}\exp(\frac{1}{2}\phi''(\omega_{sp},\theta)(\omega-\omega_{sp})^2)d\omega \\
&= \frac{E(0,\omega_{sp})}{\sqrt{2\pi}}e^{i\phi(\omega_{sp},t)}\sqrt{\frac{2\pi}{-i\phi''(\omega_{sp})}}
\end{aligned}
\tag{2.54}
$$

From the spectrum of the input pulse, the two transient fields at two types of saddle-points are given as

$$
\begin{aligned}
E_S(z,t) &= \frac{iE_0}{2\pi(\omega_{sp}^+ - \omega_p)}e^{i\phi(\omega_{sp}^+,t)}\sqrt{\frac{2\pi}{-i\phi''(\omega_{sp}^+,t)}} \\
&\approx \frac{iE_0}{\sqrt{2\pi}\Delta_{sp}^+}\frac{e^{i\phi(\omega_{sp}^+,t)-i\pi/4}}{\sqrt{\alpha_0 z\gamma/\xi(t)^3}} \approx \frac{iE_0}{\sqrt{2\pi}}\frac{e^{i(\sqrt{2\alpha_0 z\gamma\tau}-\pi/4)}}{(2\alpha_0 z\gamma\tau)^{1/4}}
\end{aligned}
\tag{2.55}
$$

$$
\begin{aligned}
E_B(z,t) &= \frac{iE_0}{2\pi(\omega_{sp}^- - \omega_p)}e^{i\phi(\omega_{sp}^-,t)}\sqrt{\frac{2\pi}{-i\phi''(\omega_{sp}^-,t)}} \\
&\approx \frac{iE_0}{\sqrt{2\pi}\Delta_{sp}^-}\frac{e^{i\phi(\omega_{sp}^-,t)+i\pi/4}}{\sqrt{\alpha_0 z\gamma/\xi(t)^3}} \approx \frac{iE_0}{\sqrt{2\pi}}\frac{e^{-i(\sqrt{2\alpha_0 z\gamma\tau}-\pi/4)}}{(2\alpha_0 z\gamma\tau)^{1/4}}
\end{aligned}
\tag{2.56}
$$

Here, the first approximation can be made because of the on-resonance condition $\omega_p \approx \omega_0$, so that $\omega_{sp}^{\pm} - \omega_p \approx \omega_{sp}^{\pm} - \omega_0 = \Delta_{sp}^{\pm}$.

The second approximation is due to the narrow-resonance condition ($\gamma \to 0$), so that $\Delta_{sp}^{\pm}(t) \approx \xi(t)$.

$$E_{SB}(z,t) = E_S(z,t) + E_B(z,t)$$

$$= \sqrt{\frac{2}{\pi\sqrt{2\alpha_0 z\gamma\tau}}} E_0 e^{-i\omega_p\tau} \cos\left(\sqrt{2\alpha_0 z\gamma\tau} - \frac{\pi}{4}\right) \quad (2.57)$$

By assuming highly absorptive media, the total transmitted field only consists of Sommerfeld and Brillouin precursors, and main signal is zero due to absorption by two-level atoms. In the next section, we will discuss EIT media where the delayed main signal can be transmitted without absorption.

From the Bessel function approximation $J_0(x) \approx \sqrt{\frac{2}{\pi x}}\cos(x - \pi/4)$, we finally obtain the total transient field as,

$$E_{SB}(z,t) = E_0 J_0(\sqrt{2\alpha_0 z\gamma\tau})\Theta(\tau)e^{-\gamma\tau}e^{(k_0 z - \omega_0\tau)} \quad (2.58)$$

The above expression is equivalent to the zeros order of Bessel term in Eq. (2.48) for on-resonance condition. One can prove that for a step-off pulse $E_0(z = 0,t) = E_0\Theta(-t)$ with falling edge, the SB precursor is the same as equation (2.58) but with a minus sign. Therefore, for step input pulse $E_0(z = 0,t) = E_0\Theta(\pm t)$, we obtain

$$E_{SB}(z,t) = \pm E_0 J_0(\sqrt{2\alpha_0 z\gamma\tau})\Theta(\tau)e^{-\gamma\tau}e^{i(k_0 z - \omega_0\tau)} \quad (2.59)$$

The interesting thing is that the identical expression derived in two approaches has been discussed over and over in the past with various impulse responses, such as 0π-pulse [13]. By solving Maxwell-Bloch equation, the analytic expression of 0π-pulse turned out to be exactly the same as Eq. (2.59). Although quite a few could notice that 0π-pulse is identical to resonant precursors [14], majority of optical community, even Crisp, had denied the existence of optical precursors in small area (0π)-pulse analysis. However, due to the experimental demonstration of resonant precursors, the analogy between optical precursors and 0π-pulse can be accepted and both phenomena are interpreted as coherent transients [15].

2.3.2 Main Signal Propagation in Electromagnetic Induced Transparency Medium

Conventional theory of optical precursors deals with a single Lorentz oscillator, as we have discussed so far. In a single Lorentz medium, the main signal corresponding to the pole $\omega \to \omega_0$ is absorbed heavily. What happens if we consider EIT medium? How does the change affect total transmitted field? We would like to answer such questions in this last section based on the Ref. [18].

For a three-level EIT system, the model of simple Lorentz oscillators does not apply. We can instead take a semi-classical approach to obtain the dielectric function. As depicted in Fig. 2.7a, the three-state system is coupled with a strong

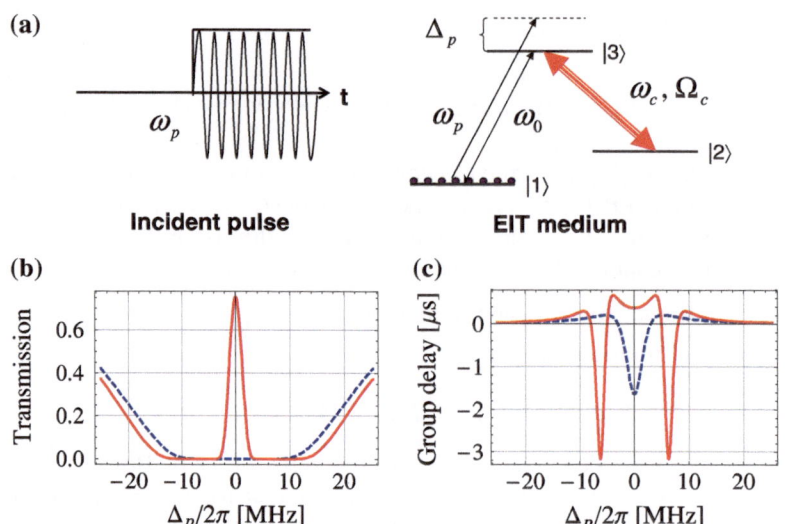

(a)

ω_p Incident pulse

Δ_p |3⟩

ω_c, Ω_c

ω_p ω_0 |2⟩

|1⟩

EIT medium

(b)

(c)

Fig. 2.7 Optical pulse propagation though an EIT. The figure was published in Jeong H and Du S (2009) Phys. Rev. A(R) 79: 011802

coupling field. Assume that the coupling field is on-resonance with the transition $|2\rangle \rightarrow |3\rangle$. The Hamiltonian in the rotating-wave wave is shown below, including the relaxation mechanism with decay rate of $|2\rangle$ ($|3\rangle$) as $\Gamma_2 = 2\gamma_{12}$ ($\Gamma_3 = 2\gamma_{13}$):

$$H_{eff} \doteq \hbar \begin{bmatrix} 0 & 0 & -\frac{1}{2}\Omega_p^* \\ 0 & -\Delta_p - \frac{i\Gamma_2}{2} & -\frac{1}{2}\Omega_c^* \\ -\frac{1}{2}\Omega_p & -\frac{1}{2}\Omega_c & -\Delta_p - \frac{i\Gamma_3}{2} \end{bmatrix} \tag{2.60}$$

where $\Delta_p = \omega_p - \omega_0$. From $|\varphi\rangle_R = a_1(t)|1\rangle + a_2(t)|2\rangle + a_3(t)|3\rangle$, the coupled differential equations:

$$i\hbar \begin{bmatrix} \dot{a}_1 \\ \dot{a}_2 \\ \dot{a}_3 \end{bmatrix} = \hbar \begin{bmatrix} 0 & 0 & -\frac{1}{2}\Omega_p^* \\ 0 & -\Delta_p - \frac{i\Gamma_2}{2} & -\frac{1}{2}\Omega_c^* \\ -\frac{1}{2}\Omega_p & -\frac{1}{2}\Omega_c & -\Delta_p - \frac{i\Gamma_3}{2} \end{bmatrix} \begin{bmatrix} a_1 \\ a_2 \\ a_3 \end{bmatrix} \tag{2.61}$$

With ground state approximation $a_1 \approx 1$, the steady state for the system considering relaxation mechanism can be obtained from the following calculation:

$$0 = \left(-\Delta_p - \frac{i\Gamma_2}{2} \right) a_2 - \frac{1}{2}\Omega_c^* a_3$$

$$0 = -\frac{1}{2}\Omega_p - \frac{1}{2}\Omega_c a_2 - \left[\Delta_p + \frac{i\Gamma_3}{2} \right] a_3 \tag{2.62}$$

Therefore, the probability amplitude of state $|3\rangle$ is:

$$a_3 = \frac{2\Omega_p(\Delta_p + \frac{i\Gamma_2}{2})}{|\Omega_c|^2 - 4(\Delta_p + \frac{i\Gamma_3}{2})(\delta + \frac{i\Gamma_2}{2})} \tag{2.63}$$

The induced polarization density is:

$$\begin{aligned} P &= \langle \varphi | \hat{P} | \varphi \rangle \\ &= N(a_1^* a_3 \mu_{13} e^{-i\omega_p t} + a_2^* a_3 \mu_{23} e^{-i\omega_c t} + a_1 a_3^* \mu_{13}^* e^{i\omega_p t} + a_2 a_3^* \mu_{23}^* e^{i\omega_c t}) \end{aligned} \tag{2.64}$$

If only the polarization induced by the probe laser beam is considered, Eq. (2.62) becomes:

$$P = N(a_1^* a_3 \mu_{13} e^{-i\omega_p t} + a_1 a_3^* \mu_{13}^* e^{i\omega_p t}) \tag{2.65}$$

Therefore, we obtain the linear susceptibility χ of this three-level system:

$$\begin{aligned} \chi(\omega) &= \frac{2N\mu_{13}a_3}{\varepsilon_0 E_p} = \frac{N|\mu_{13}|^2}{\varepsilon_0 \hbar} \cdot \frac{4(\Delta_p + \frac{i\Gamma_2}{2})}{|\Omega_c|^2 - 4(\Delta_p + \frac{i\Gamma_3}{2})(\Delta_p + \frac{i\Gamma_2}{2})} \\ &= \frac{\alpha_0 z}{k_p z} \cdot \frac{2\Gamma_3(\Delta_p + \frac{i\Gamma_2}{2})}{|\Omega_c|^2 - 4(\Delta_p + \frac{i\Gamma_3}{2})(\Delta_p + \frac{i\Gamma_2}{2})} \end{aligned} \tag{2.66}$$

For EIT medium, with high optical depth condition $\alpha_0 z \gg 1$, the saddle points are far-detuned from the resonance, we can treat the SB precursor field the same as a two-level system. The analytic expression of the SB precursors is same as Eqs. (2.58–2.59). The big difference comes from the main signal part which is not absorbed within the EIT window. To describe the main signal, the above stationary phase approximation (2.50) is not appropriate because $\phi''(\omega_c) = 0$. To obtain the main field expression, we could deal with the impulse response (Green's function) of the EIT window $[(\omega_0 - \Delta_e, \omega_0 + \Delta_e)]$,

$$G_{EIT}(z, t) = \frac{1}{2\pi} \int_{\omega_0 - \Delta_e}^{\omega_0 + \Delta_e} e^{i[k(\omega) - \omega t]} d\omega \tag{2.67}$$

where $\Delta_e \equiv \Omega_c/2$. As we convolute the input signal $E(0, t)$ with Green's function $G_{EIT}(z, t)$, the main signal $E_C(z, t)$ can be expressed as

$$\begin{aligned} E_C(z, t) &= \frac{1}{2\pi} \int_{-\infty}^{\infty} G_{EIT}(z, t - \tau) E_0(0, \tau) d\omega \\ &= \int_0^{\infty} G_{EIT}(z, t - \tau) d\omega \end{aligned} \tag{2.68}$$

The interesting point is that Eq. (2.68) implies active control of main signal by varying the parameters of the medium and control beam.

With reasonable approximation, the analytical expression of the main field is obtained as follows [19]. The ground-state dephasing is negligible, i.e., $\gamma_{12} \approx 0$ or

$|\Omega_c|^2 \gg 4\gamma_{12}\gamma_{13}$. The carrier frequency of the probe beam is ω_p, while we now denote ω as the frequency deviation from ω_p due to the spectral constitutes of the finite probe pulse. Since the EIT transparency window is very narrow, the EIT linear susceptibility can be expanded to the third order near $\omega = 0$:

$$
\begin{aligned}
\chi(\omega) &\approx \chi(0) + \chi'(0)\omega + \frac{1}{2}\chi''(0)\omega^2 + \frac{1}{6}\chi'''(0)\omega^3 \\
&= \frac{4i\alpha_0\gamma_{12}\gamma_{13}}{k_{p0}\Omega_c^2} + \frac{4\alpha_0\gamma_{13}}{k_{p0}\Omega_c^2}\omega + \frac{\omega^2}{2}\frac{32i\alpha_0\gamma_{13}^2}{k_{p0}\Omega_c^4} + \frac{\omega^3}{6}\frac{12\alpha_0\gamma_{13}(-\frac{32\gamma_{13}^2}{\Omega_c^6}+\frac{8}{\Omega_c^4})}{k_{p0}}
\end{aligned}
$$
(2.69)

The transfer function can then be approximated as:

$$
T(\omega) = e^{-\gamma_{12}\tau_g}e^{-\omega^2/(2a^2)}e^{i\omega\tau_g}e^{i\omega^3/(3b^3)}
$$
(2.70)

where $\tau_g = 2\alpha_0 z\gamma_{13}/|\Omega_c|^2$ is the group delay, $a = \sqrt{\alpha_0 z}/(2\tau_g)$ determines the EIT bandwidth, $b = |\Omega_c|^2[24\alpha_0 z\gamma_{13}(|\Omega_c|^2-4\gamma_{13}^2)]^{-1/3}$. The impulse response function can be expressed as a convolution:

$$
h(t) = \frac{ab}{\sqrt{2\pi}}e^{-\gamma_{12}\tau_g}e^{-\frac{1}{2}a^2(t-\tau_g)^2} * Ai(-bt)
$$
(2.71)

$Ai(bt)$ is an Airy function, which comes from the third-order term of the linear susceptibility. The main field is the convolution of the input step pulse with the impulse response function:

$$
\begin{aligned}
E_{M\pm}(t) &= E_0\Theta(\pm t) * h(t) \\
&= \frac{E_0}{2}e^{-\gamma_{12}\tau_g}(1 \pm erf[\frac{a(t-\tau_g)}{\sqrt{2}}]) * bAi(-bt)
\end{aligned}
$$
(2.72)

where *erf* denotes the error function, which indicates that the main field is delayed by τ_g. The Airy function adds a small modulation on top of the main field and leads to the "postcursor" introduced in Mache and Segard's work in 2009. But in most cases, the Airy function effect is small and can be ignored.

References

1. Brillouin, L.: Wave Propagation and Group Velocity. Academic Press, New York (1960)
2. Oughstun, K.E., Sherman, G.C.: Electromagnetic Pulse Propagation in Causal Dielectrics. Springer, Berlin (1994)
3. Jeong, H., Dawes, A.M., Gauthier, D.J.: Direct observation of optical precursors in a region of anomalous dispersion. Phys. Rev. Lett. **96**, 143901 (2006)
4. Wei, D., Chen, J.F., Loy, M.M.T., Wong, G.K.L., Du, S.: Optical precursors with electromagnetically-induced transparency in cold atoms. Phys. Rev. Letts. **103**, 093602 (2009)

5. Jeong, H., Österberg, U.L., Hansson, T.: Evolution of Sommerfeld and Brillouin precursors in intermediate spectral regimes. JOSA B **26**, 2455–2460 (2009)
6. LeFew, W.R., Venakides, S., Gauthier, D.J.: Accurate description of optical precursors and their relation to weak-field coherent optical transients. Phys. Rev. A **79**, 063842 (2009)
7. Jeong, H., Du, S.: Slow-light-induced interference with stacked optical precursors for square input pulses. Opt. Lett. **35**, 124–126 (2010)
8. Jeong, H., Österberg, U.L.: Steady-state pulse component in ultrafast pulse propagation in an anomalously dispersive dielectric. Phys. Rev. A **77**, 021803 (2008)
9. Jeong, H.: Direct observation of optical precursors in a cold potassium gas. Ph.D. Dissertation (unpublished) (2006)
10. Copson, E.T.: Asymptotic Expansions. University Press, Cambridge (1965)
11. LeFew, W.R.: Optical precursor behavior. Ph.D. Dissertation, Duke University (unpublishd) (2007)
12. Jeong, H.J., Dawes, A.M.C., Gauthier, D.J.: Carrier-frequency dependence of a step-modulated pulse propagating through a weakly dispersive single narrow-resonance absorber. J. Mod. Opt. **58**(10), 865–872 (2011)
13. Crisp, M.D.: Propagation of small-area pulses of coherent light through a resonant medium. Phys. Rev. A **1**, 1604–1611 (1970)
14. Varoquaux, E., Williams, G.A., Avenel, O.: Pulse propagation in a resonant medium: Application to sound waves in superfluid ^{3}He. Phys. Rev. B **34**, 7617–7640 (1986)
15. Jeong, H., Österberg, U.: Coherent transients: Optical precursors and 0π pulses. JOSA B **25**, B1–B5 (2008)
16. Aaviksoo, J., Lippmaa, J., Kuhl, J.: Observability of optical precursors. JOSA B **5**, 1631–1635 (1988)
17. Arfken, G.B., Weber, H.J., Ruby, L.: Mathematical Methods for Physicists, vol. 3, pp. 428–431. Academic Press, San Diego (1985)
18. Jeong, H., Du, S.: Two-way transparency in the light-matter interaction: Optical precursors with electromagnetically induced transparency. Phys. Rev. A **79**, 011802 (2009)
19. Chen, J.F., Wang, S., Wei, D., Loy, M.M.T., Wong, G.K.L., Du, S.: Optical coherent transients in cold atoms: From free-induction decay to optical precursors. Phys. Rev. A **81**, 033844 (2010)

Chapter 3
Searching for Precursors: From Microwave to Primary Optical Experiments

Abstract In this chapter, we review some early experimental works on precursor observation. The first experimental observation of Sommerfeld-Brillouin precursor is reported in coaxial transmission line, where SB precursors exist in microwave. Later, precursors are studied and measured experimentally in sound-wave domain, with superfluid ^{3}He-B prepared closer to "resonant regime". The first optical precursor was reported in infrared light propagated through GaAs single-crystal layer. We review and discuss these experimental works within the theoretical framework in Chap. 2.

Theoretical calculation on precursor phenomena was first made by Sommerfeld and Brillouin in 1914, with a purpose to looking for maximum velocity of a signal carried by optical pulse. In the previous chapters, our discussion focused on the asymptotic approach we used to obtain the SB precursor field. In this chapter, we would like to review our experimental observations of the transient signals in microwave, sound, and optical domain. With a dispersive medium model with broad resonance, SB predicted that the precursor field is an extremely rapid transient with femto-second time-scale profile, and with almost negligible magnitude. Oughstun and Sherman later recalculated the problem and verified the availability of measurement. However, the dispersive medium, indeed usually with broad resonance, puts the coherence time of the transient beyond the measurement limit. There were no reports on precursor observation in electromagnetic waves, including microwaves and optical waves, until 1969, when Pleshko and Palócz [1] reported the first microwave experiment with a transmission waveguide. In this artificially controlled dispersive medium, the plasma frequency can be tuned to be appropriate for the precursor in microwave with sub-nano to nano second scale.

JF Chen

JF Chen et al., *Optical Precursors*, SpringerBriefs in Physics,
DOI: 10.1007/978-981-4451-94-9_3, © The Author(s) 2013

3.1 In Microwave Frequency Domain: First Precursor

The first experimental observation of precursor was reported by Pleshko and Palócz [1] in the microwave frequency domain. The fabricated medium was a coaxial transmission waveguide. In this fabricated waveguide, the resonant line was located at the frequency of the order of GHz. Therefore the dispersion curve at high and low frequencies was not symmetric with respect to the resonant spectral line. The precursor at the first arrival time $t_0 = z/c$ comes from the extremely large frequencies $\omega \to \infty$. The refraction index at large frequencies is approximated as:

$$n(\omega) \approx 1 - \frac{\omega_{pl}^2}{2\omega^2} \qquad (3.1)$$

From Eqs. (2.12), (2.15) and (3.1), the amplitude is derived as:

$$E(z,t) = \frac{E_0}{2\pi i} \int \frac{d\omega}{\omega - \omega_p} e^{-i[(t-t_0)\omega + \xi/\omega]} \qquad (3.2)$$

in which the contour is taken in a counterclockwise direction around the whole circle with a large radius R, and $\xi = \frac{\omega_{pl}^2 z}{2c}$. Change the variable $v = -i(t - t_0)\,\omega$ in Eq. (3.2), with limits $\omega \gg \omega_p$, and we could transform Eq. (3.2) into

$$E(z,t) = \frac{E_0}{2\pi i} \int \frac{dv}{v} e^{v - \frac{\xi(t-t_0)}{v}} \qquad (3.3)$$

Note that Bessel function with integer m has a form as:

$$J_m(x) = \frac{1}{2\pi i} \left(\frac{x}{2}\right)^m \int \frac{dv}{v^{m+1}} e^{v - (x^2/4v)} \qquad (3.4)$$

One can identify that if $x = 2\sqrt{\xi(t - t_0)}$, Eq. (3.5) becomes,

$$E(z,t) = E_0 J_0[2\sqrt{\xi(t - t_0)}] \qquad (3.5)$$

Equation (3.5) indicates that the high-frequency components of the incident wave arrive at the earliest with velocity close to c, and they give rise to Sommerfeld precursor expressed as Bessel function. The oscillation frequency of Sommerfeld precursor increases with parameter ξ, or say, with plasma frequency of the medium. According to Sommerfeld and Brillouin's theory, a typical value for plasma frequency is $\omega_{pl} = 10^{16}\,\text{s}^{-1}$, and thus the oscillating period of Sommerfeld precursor is of the order of $10^{-20}\,\text{s}$. Hence, with these parameters, Sommerfeld precursor is indiscernible with the present detection technique. However, in the waveguide experiments, the plasma frequency of the medium is of the order of 10^9, and the precursor oscillates with sub-nano-second to nano-second. Figure 3.1 shows the high frequency limit case, with $\xi = 10^{11}$ in simulation.

Fig. 3.1 Theoretical simulation based on the Sommerfeld precursor in waveguide at the high frequency limit

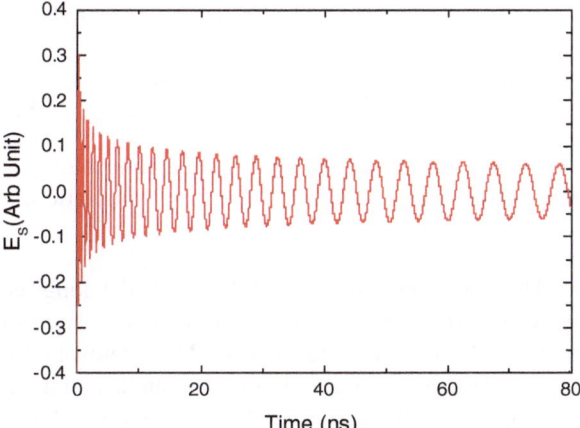

On the low frequency side in the spectrum, the stationary phase point $\omega \to 0$ contributes to the Brillouin precursor, which arrives at $t_1 = \frac{n(\omega=0)z}{c}$. At this stationary phase point, ϕ' and ϕ'' vanish, and we keep terms up to ω^3 to expand $\phi(\omega)$:

$$\phi(\omega) = -\omega(t - t_1) + \frac{z}{6}\left(\frac{d^3k}{d\omega^3}\right)_0\omega^3 \tag{3.6}$$

The transmitted amplitude is written as,

$$E(z, t) = \int E_0(\omega)d\omega e^{i[(t_1-t)\omega+(z/6)(d^3k/d\omega^3)_0\omega^3]} \tag{3.7}$$

Here $E_0(\omega)$ represents the spectrum of the incident pulse, and for convenience, in low frequency limits we consider it to be constant. From Eq. (3.1) the third-order derivative becomes,

$$\frac{dk^3}{d^3\omega}\Big|_0 = \frac{3\omega_{pl}^2}{\omega_0^4 c\sqrt{1 + \frac{\omega_{pl}^2}{\omega_0^2}}} = \frac{3\omega_{pl}^2}{\omega_0^4 cn(0)} \tag{3.8}$$

The integration in Eq. (3.7) can be written as an Airy function, if we change the variables as [2]:

$$t' = \left(\frac{2}{z(d^3k/d\omega^3)_0}\right)^{1/3}(t_1 - t) \tag{3.9}$$

$$\omega' = \frac{1}{2\pi}\left(\frac{z(d^3k/d\omega^3)_0}{2}\right)^{1/3}\omega \tag{3.10}$$

Therefore the Brillouin precursor is expressed as Airy function:

$$E_B(z,t) = 2\pi \left(\frac{2}{z(d^3k/d\omega^3)_0} \right)^{1/3} Ai\left((t_1 - t) \left(\frac{2}{z(d^3k/d\omega^3)_0} \right)^{1/3} \right) \qquad (3.11)$$

Figure 3.2 shows the Airy function contributing to the Brillouin precursor from low frequency limit. In the simulation, we set the parameter $\left(\frac{2}{z(d^3k/d\omega^3)_0} \right)^{1/3} = 2 \times 10^8$.

The waveguide was first filled with the longitudinally magnetized ferromagnetic material, and the dispersion characteristic could be varied by the applied magnetic field. The incident wave was a sinusoidal wave train with a frequency of 0.625 GHz, in microwave frequency range, and the electric field could be captured directly by a high-bandwidth oscilloscope. After varying the magnetic field applied onto the waveguide, the plasma frequency was well controlled. As shown in Fig. 3.3 from Ref. [1], when the magnetic field was low, the Sommerfeld precursors were superimposed on the Brillouin precursor. When the magnetic field increased and the plasma frequency grew, Sommerfeld precursor became invisible since the oscillation was beyond the detection limit. To observe the Sommerfeld and Brillouin precursor separately, they further selected two different guiding structures to mimic the two extreme frequency branches. High-bandwidth function generator (able to produce rise time as short as 30 ps) and high speed oscilloscope (12.4 GHz sampling oscilloscope) was utilized in the direct measurement of the electric field. Figure 3.4 shows the case for precursors at low frequency branch, which matches our previous simulation at the frequency extreme (as shown in Fig. 3.2). At the other frequency extreme, air filled waveguide was utilized and the Sommerfeld precursor was claimed to be observed with the C-band waveguide with cut-off frequency of 4.29 GHz. However, the frequency components and the delay time of the transient signal is evaluated from the oscilloscope with a rather

Fig. 3.2 Theoretical simulation based on the Brillouin precursors in waveguide at the low frequency limit

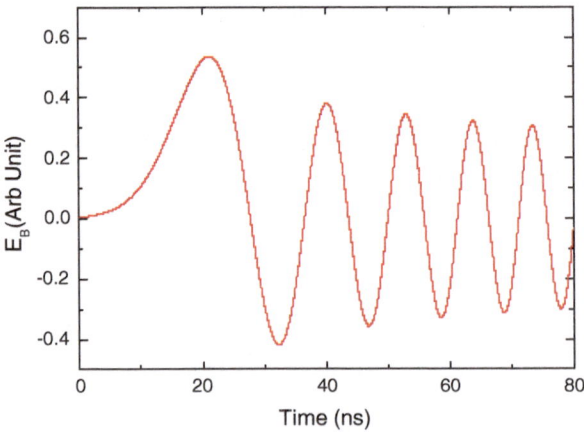

Fig. 3.3 Output waveform in microwave domain, with different plasma frequency controlled by the applied magnetic field. The scale of the measurement is of 5 ns/ div. **a** 20 GAUSS, **b** 100 GAUSS. Reprinted Fig. 2 with permission from Pleshko and Palócz [1]. Copyright 2013 by the American Physical Society

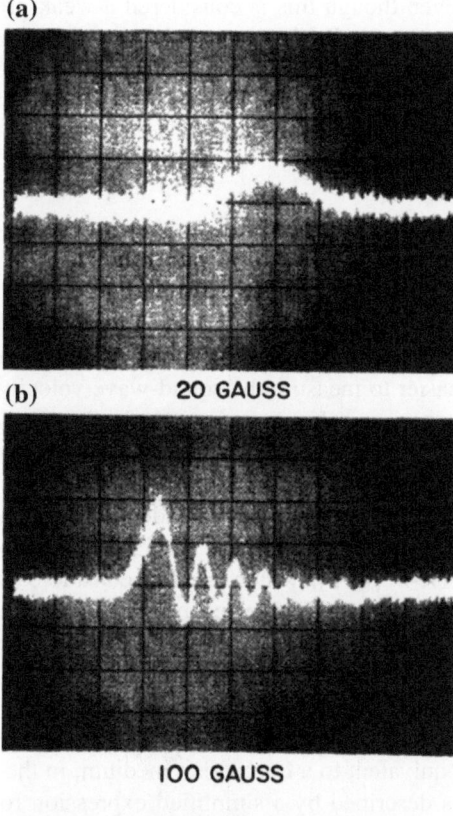

(a)

(b)   20 GAUSS

IOO GAUSS

Fig. 3.4 Output waveform measured in material of RG8/ U *coaxial line*. Reprinted Fig. 3 with permission from Pleshko and Palócz [1]. Copyright 2013 by the American Physical Society

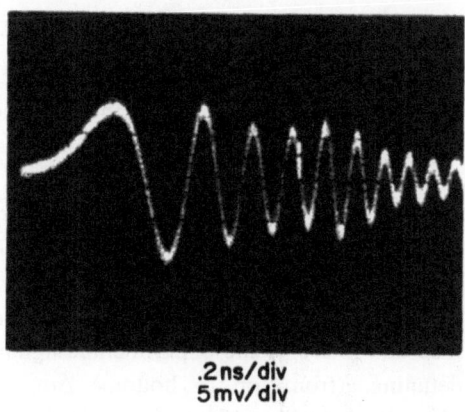

.2 ns/div
5 mv/div

rough approach, and we believe that the observed transient signal needs further verification with modern advanced technique.

In conclusion, the first experimental observation of SB precursor was carried out in microwave regime, with controllable energy band of waveguide medium.

Even though this is considered a weak proof for Sommerfeld precursor, the work shows the Brillouin precursors with considerable magnitude. This is the unique attempt to demonstrate the precursor field through the real incident field. In contrast, precursors carried out in optical domain are depicted in field amplitude envelope, due to the high oscillation (>10 THz) with optical frequency.

3.2 Observation of Sound Wave with Superfluid ^{3}He-B

When the electromagnetic waves propagate with a speed in the order of physical constant c, sound waves travel much slower with ~ 340 m/s. Therefore it is much easier to measure the sound-wave velocity with satisfactory accuracy. It is a good experimental approach to study the problem of wave propagation, especially on phase velocity, group velocity and information velocity. Further, sound wave is classified as a mechanical wave, different from electromagnetic wave. Precursors in sound-wave realm provide reference for comparison with wave propagation in other realms. Avenel et al. (1983) [2] reported sound-wave precursor in superfluid ^{3}He-B [2, 3], which can be considered an ideal homogeneous medium at low temperature. The experiments were carried out in a copper nuclear demagnetization cryostat. The sound transducers operated at constant frequency and the mode resonance was temperature dependent. In such a special medium, the sound wave was propagated with very weak damping. Further, the wave was propagating with a carrier frequency close to the resonant frequency of the superfluid, which was equivalent to a Lorentzian medium. In the resonant regime, the medium dispersion is described by a simplified expression for $k(\omega)$, according to Eq. (2.30),

$$k(\omega) = \frac{\omega}{c} - \frac{\omega_{pl}^2}{\omega - \omega_0 + i\gamma} \tag{3.12}$$

With slowly varying approximation, the precursor field can be described with Bessel function,

$$E_{SB}(z,t) = E_0 J_0(\sqrt{2\alpha_0 z\gamma\tau})\Theta(\tau)e^{-\gamma\tau}e^{(k_0 z - \omega_0 \tau)} \tag{3.13}$$

where γ denotes the decay coefficient. The typical parameters in Eq. (3.12) are: $\lambda = 3.7\,\mu$m, velocity for sound wave propagation in the sonic cell $c_0 = 202$ m/s, $\omega_0 = 55$ MHz and $\gamma = 90 \times 10^3$ rad/s. According to Varoquaus et al., the amplitude and phase of the experimental signal are shown in Fig. 3.2, with excitation detuning, (from top to bottom) $\Delta\omega = 9.4 \times 10^5$ rad/s, 4.3×10^5 rad/s, 2.4×10^5 rad/s and 6.4×10^4 rad/s. The least-detuning case (bottom curve in Fig. 3.2) exhibits clear wiggles demonstrating Bessel function of Eq. (3.13). In such an acoustic system, the electronic signal characteristic of rectangular envelope must be transformed to a mechanical signal through a transducer. The rectangular-envelope hence undergoes a convolution with the response function $H(s)$ of the transducer. The actual envelope of the incident sound wave is given by,

$$A(s) = H(s)A_{rf}(s) \tag{3.14}$$

Once more, at the receiver end, the mechanical signal is transformed to the electronic signal. Therefore the sharp fronts of the rectangular envelope are smoothed by the transformation, before emitting into the liquid medium. The sound wave again is transformed to electronic signal at the receiver end. The top curve in Fig. 3.5 shows the envelope received at the electronic device without experiencing dispersion: $A(s) = H^2(s)A_{rf}(s)$.

Fig. 3.5 Sound-wave precursor reported in Varoquaux et al. [3]. The experimental results are plotted with dots and the calculated curves are plotted with solid line. Reprinted Fig. 1 with permission from Avenel et al. [2]. Copyright 2013 by the American Physical Society

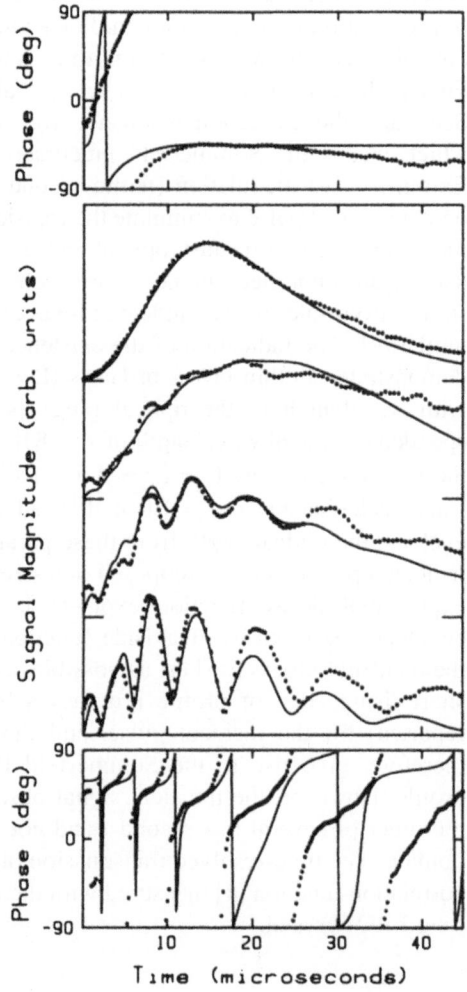

Superfluid in ^{3}He-B was explored to be the medium for observing the precursor propagation, due to low-dissipation and close to resonance. However, because of the relatively slow response of mechanical wave, rectangular envelope of sound wave is hard to be generated in experiments. Therefore the SB precursor aroused by the sudden change of incident field was not separate from the main wave train in the measurement. In contrast, in optical regime, this problem is easily solved with well controlled optical pulses.

3.3 Searching for Precursors in Optical Domain: In GaAs Crystal and Water

When Sommerfeld and Brillouin introduced precursors, they had in time problem of propagation in optical domain. For a long time, to demonstrate the existence of optical precursors was considered to be a hard work. There are two main reasons. Firstly, the carrier frequency of an optical pulse is extremely high ($\sim 10^{14}$ Hz). Secondly, the excitation resonance for common medium has broad spectrum, which means the Sommerfeld precursor must be having ultra-high frequency. Extremely fast modulation (femto-second to sub-pico-second) should be applied onto an optical pulse to stimulate the transient signals. Technology on optical pulse modulation and ultra-fast optical pulse detection is still being developed and usually does not meet the necessary requirements. Therefore, experimental works started from microwave, and sound-wave, and were not reported in optical range until 1991. The indication of the existence of precursors in optical range was first demonstrated in thin layers of GaAs (L = 0.2 μm), with ultrashort optical pulses with wavelength in the optical range [4]. The effective excitation line corresponded to optical wavelength of $\lambda = 818.3$ nm, with full width half maximum of the resonance-line as $\Gamma = 2\gamma = 7.5 \times 10^{11}$ rad/s. The effective plasma frequency can be calculated as $\omega_{pl} = 1.0 \times 10^{14}$ rad, and so that $\omega_{pl} \sim 0.1\omega_0$. Also, we could evaluate the optical depth from these parameters, with z = 0.2 μm, $\alpha_0 L \approx 9$. The incident optical pulse is composed of a steep rising edge followed by a single-sided exponential decay function $\exp(-t/\tau)$. To approach the ideal step-on pulse envelope described by Heaviside function $E_{0+}(t) = E_0\Theta(+t)$, the rising edge of the incident pulse is as short as possible, and it is 0.5 ps in the experiment. Such a short rising edge of profile provides a broad range spectrum ($\sim 10^{12}$ rad/s as reported) for the incident pulse, and those far-detuned frequency components, therefore, give rise to the Sommerfeld Brillouin precursor in the transmission profile. However, the transient signal observed in the GaAs layer is very fast, of the order of several picoseconds, and not able to be detected directly in the time domain. The time-resolved transmission of the optical pulse was attained via cross correlation function [4] measured with the initial laser pulse in a several millimeter long LiIO$_3$ crystal.

The condition $\omega_{pl} \ll \sqrt{8\omega_0\gamma}$ is not satisfied in this system, and thus Eq. (2.7) should be dealt without approximation. It is not easy to gain an analytical solution for such parameter regime described above. However, with refractive index feature displayed as Eq. (2.13),

$$n(\omega) = \sqrt{1 - \frac{\omega_{pl}^2}{\omega^2 - \omega_0^2 + 2i\omega\gamma}} \tag{3.15}$$

one could obtain a numerical result from frequency-domain linear dispersion theory. For sufficiently weak excitation, the atomic population remains mostly in the ground state, and the Fourier frequency components of the incident pulse can be treated independently and nonlinear wave mixing between them can be ignored with first-order perturbation theory. The induced electric dipole in the frequency domain at the transition is determined by the linear relation:

$$P(\omega) = \varepsilon_0 \chi(\omega) E(\omega) \tag{3.16}$$

where the linear susceptibility of the dispersive medium is denoted as $\chi(\omega)$. Next, we assume that, every frequency component of the input pulse is evolving within the medium independently and is governed by corresponding linear transfer function. The propagation of the weak probe field envelope is given by the integral

$$E(L,t) = \frac{1}{2\pi} \int E_0(\omega) e^{i[\Delta k(\omega)z - \omega t]} d\omega \tag{3.17}$$

in which we make an approximation $\sqrt{1 + \chi} = 1 + \frac{1}{2}\chi$, and thus $\Delta k(\omega) = \frac{\omega_p}{2c}\chi(\omega)$. The linear transfer function can be expressed as,

$$T(\omega) = e^{i\Delta k(\omega)z} \tag{3.18}$$

Equation (3.16) can be expressed as the transfer function:

$$E(L,t) = \frac{1}{2\pi} \int E_0(\omega) T(\omega) e^{-i\omega t} d\omega \tag{3.19}$$

Therefore, with Fourier transform of input probe pulse $E_0(t)$, the transmitted field can be readily obtained from the integral (3.19). Figure 3.6 shows the CCF signal with detuned wave number $\Delta = 3.1\,\mathrm{cm}^{-1}$, equivalent to angular frequency 5.8×10^{11} rad/s. As shown in the figure, the peak at the very front is SB precursor and the main signal with group velocity becomes notable because of the incident exponential decay pulse shape. Normally the main signal is almost completely absorbed through the medium. However, with optical pulse absorption and reemission appropriately excited by the input exponential waveform, the transmitted pulse exhibits precursor followed by delayed main signal within the same intensity envelope.

On the other hand, Choi and Österberg [5] reported the propagation of optical precursor in distilled water. Instead of sending a short pulse with steep rising or

Fig. 3.6 CCF intensity
measurement of transmitted
optical pulse through GaAs
thin layer. Reprinted Fig. 3
with permission from
Aaviksoo et al. [4] Copyright
2013 by the American
Physical Society

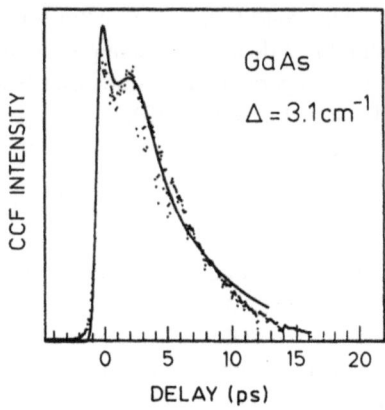

falling edges, they coupled the laser source into a holey fiber to generate a broad
bandwidth for the optical pulse. The incident pulse was ~ 540 fs long (with a
bandwidth of 60 nm), with linear chirp applied. Their evidence for the existence of
precursor is reflected in two points as described below. Firstly, the Brillouin
precursor from low frequencies in the pulse varies as $1/\sqrt{z}$ with distance when
$z \to \infty$, according to asymptotic approximation. Choi and Österberg [5] compared
the exponential decay of main pulse $e^{-\alpha z}$ with the observed attenuation in the
measurement, and concluded that the behavior of the observed peak matched the
calculation for Brillouin precursor. Secondly, at large propagation depth
$z > 500$ mm, the optical pulse broke up and the lagging peak seemed to be
growing with distance with respect to the main pulse peak. However, the experimental work was questioned. Roberts [6] introduced his theoretical work and
pointed out that non-exponential decay is only valid for pulses with dc content.
Österberg [7] argued that the theory developed by Roberts failed to describe the
behavior of transient signal such as precursor. However, he admitted that it was
difficult to determine the distance dependence of the amplitude of Brillouin pre-
cursor approximated as $\frac{1}{[\varphi''(\omega_{sp})z]^{1/2}} e^{iz\varphi(\omega_{sp})}$, since the saddle points are complicated
in the case of water. Not much later, Alfano et al. [8] debated that the pulse
breakup observed by CO did not arise from precursor but from vibration overtone
absorption and dispersion in water. By measuring the absorption spectrum of water
from 650 to 850 nm, they verified the existence of a vibrational overtone
absorption band centered at 760 nm. Also, the temporal profile of the incident
pulse was split by the band after traveling a distance of 1.2 m in water.

To conclude this section, the earlier attempts in GaAs and water are still below
what is required to compare with the analytical solutions. The absorption and
dispersion character of water is too complicated to give a clear verification for the
existence of precursor. On the other hand, for GaAs, the line width of the resonant-
line is too broad, when compared to the carrier frequency of optical pulses. The
life time for the coherent transient is too short, only being several picoseconds, to

be detected directly with time-domain signal-collection equipment. Again, the precursor at the front mingles with the main signal, without a clear signature for the separated signal. To stop the debate, the precursor community is calling for more solid evidence, with optical precursors which are able to be detected directly in time domain, and also clearly separated from the main pulse. Next round of attempts start from the absorptive medium with narrow resonance. Naturally, cold atomic ensembles become one of best candidates for the experimental measurements, with the modern technique of laser cooling and trapping.

References

1. Pleshko, P., Palócz, I.: Experimental observation of Sommerfeld and Brillouin precursors in the microwave domain. Phys. Rev. Lett. **22**, 1201 (1969)
2. Avenel, O., Rouff, M., Varoquaux, E., Williams, G.A.: Resonant pulse propagation of sound in superfluid ^{3}He-*B*. Phys. Rev. Lett. **50**(20), 1591–1594 (1983)
3. Varoquaux, E., Williams, G.A., Avenel, O.: Pulse propagation in a resonant medium: Application to sound waves in superfluid ^{3}He-*B*. Phys. Rev. B **34**, 7617–7640 (1986)
4. Aaviksoo, J., Kuhl, J., Ploog, K.: Observation of optical precursors at pulse propagation in GaAs. Phys. Rev. A **44**, R5353 (1991)
5. Choi, S.H., Österberg, U.: Observation of optical precursors in water. Phys. Rev. Letts. **92**, 193903 (2004)
6. Roberts, T.M.: Comments on "observation of optical precursors in water". Phys. Rev. Letts. **93**, 269401 (2004)
7. Gibson, U.J., Österberg, U.L.: Optical precursors and Beer's law violations; non-exponential propagation losses in water. Opt. Express **13**, 2105–2110 (2005)
8. Alfano, R.R., Birman, J.L., Ni, X., Alrubaiee, M., Das, B.B.: Comment on "observation of optical precursors in water". Phys. Rev. Letts. **94**, 239401 (2005)

Chapter 4
Observation of Optical Precursors in Cold Atoms

Abstract Cold atomic source was first introduced into the community of precursor in 2006, when Jeong (Phys. Rev. Lett. 96:143901, 2006) reported the direct observation in a cloud of potassium cold atoms. Later, Wei et al. reported observation in a 2-dimensional magneto-optical trap with optical depth as high as 50, with the assistance of EIT effect. With the advent of cold atom traps and tunable diode lasers, we now have a single physical system with parameters that can be widely tuned to cover both physical regimes. Also, in the chapter, we discuss and interpret the slow and fast light phenomena while comparing with the precursor propagation in the same system. In the last section, we review the stacked precursors measured with a multiple-step function.

Broad resonance of excitation for dispersive medium is the main problem among all the experimental works in microwave, sound, and optical waves discussed in the previous chapter. The result is that the coherence time of the transient signal is too short for direct detection in time. Also, the theoretical calculations from Oughstern and Sherman (1988) imply that resonant probing $\omega_p \approx \omega_0$ increases the magnitude of precursor to be comparable with the main signal. However, for a medium with broad resonance, the excitation frequency must be detuned from the resonant frequency to avoid heavy absorption. Motivated by the advantage of narrow-resonance medium, cold atomic ensemble was recently introduced into the optical transient community. With laser cooling and trapping technology, atoms are cooled down to the order of 0.0001 K, which is below the lowest temperature that any refrigerator could reach. Therefore, effects caused by thermal motion are depressed. Doppler broadening and atomic collision are suppressed to be negligible in cold atomic ensemble. Linewidths of atomic energy levels are only limited by natural broadening. With cold atoms as propagation medium, the single-resonance Lorentzian curve for the two-level atomic system has a narrow-resonance spectrum, characterized by $\gamma \leq \omega_0$. For example, for rubidium, the decoherence rate $\gamma = 2\pi \times 3$ MHz and hence the lifetime for optical transient signal is about 26 ns. Based on the advantages, cold atoms easily bring the optical precursors into the time-domain oscilloscope in laboratory. In cold atomic medium, the signature of precursor is the

JF Chen et al., *Optical Precursors*, SpringerBriefs in Physics,
DOI: 10.1007/978-981-4451-94-9_4, © The Author(s) 2013

clearest one compared to the observed signal in previous experiments. Furthermore, with electromagnetically-induced transparency effect, precursors are demonstrated to be prevailing the main signal. More detailed features of the transients are presented in this chapter.

4.1 Precursors in a Two-Level System

The first observation of optical precursor in cold atom source was reported by Jeong et al. [1], with a cloud of cold potassium (^{39}K) atoms cooled and trapped in a magneto-optical trap. Later, a series of observations were reported in cold rubidium (^{85}Rb) magneto-optical trap [2], with versatile experimental parameters for optical depth and excitation detuning. Figure 4.1 demonstrates the general schemes for measurement. As Fig. 4.1a shows, a weak step-modulated incident laser pulse generated via a modulator is sent through the cold atom cloud and received by PMT detector. In the following sections, we are going to consider the problem dealing with conditions of small and large optical thickness, separately. Optical thickness is characterized by optical depth ($\alpha_0 z$), which quantifies the ability of medium to absorb light. It is an important parameter for the precursor measurement.

4.1.1 In Cold Atoms with Small Optical Thickness

At condition $\alpha_0 z \sim 1$, the measurement within cold potassium (^{39}K) atoms is shown in Fig. 4.2. For off-resonant carrier frequency, $\Delta \equiv \omega_p - \omega_0 \neq 0$, the steady state signal transmits through the dielectric medium with little absorption, and the modulation frequency is inversely related to the detuning of the carrier frequency. Thus, the carrier frequency is close to the medium resonance, and the modulation of the envelope disappears. The steady-state transmission always follows the Beer-Lambert's law of absorption, so, it determines the level of the

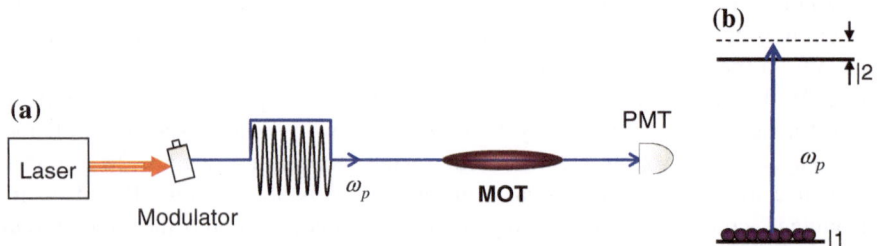

Fig. 4.1 Schemes for precursor observation in magneto-optical trap (*MOT*) with two-level energy structure. **a** General experimental setup. **b** Energy level structure

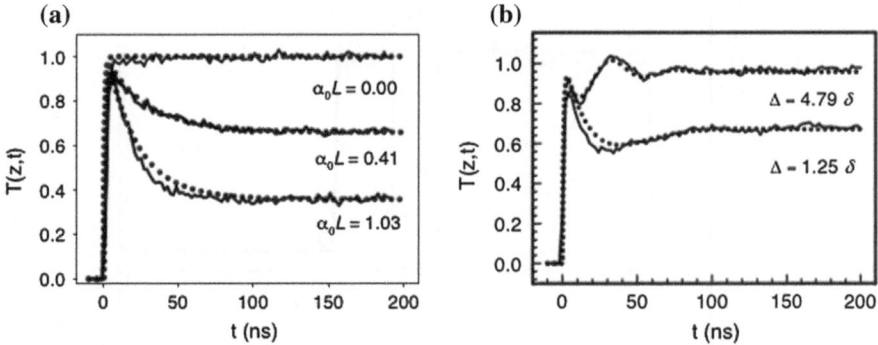

Fig. 4.2 Experimentally observed transient transmission (*black solid lines*) taken at (**a**) on-resonance ($\Delta = 0$) for three different optical depths, (**b**) for off-resonances, $\Delta \sim 5\gamma$, and $\Delta \sim \gamma$, which are compared with theory Eqs. (2.48) and (2.49) (*black dotted lines*). The figure is from Jeong thesis (2006): 22–23

transmission after the initial transients. At slightly higher optical depth $\alpha_0 z \sim 2$ reported by Chen et al. [3], the experimental results agree with the above general feature. As shown by Fig. 4.3c, for off-resonance case $\Delta = 2\pi \times 20$ MHz, the oscillation signal appears at the rising edge with a period determined by frequency detuning is $2\pi/\Delta = 50$ ns. At the falling edge, the signal intensity is proportional to $|\Delta + i\gamma|^{-2}$, and therefore the signal is reduced to almost zero and is undetected. Actually, as discussed in Ref. [3], the optical transient observed at small optical thickness condition is classified as free-induction decay (FID). Instead of oscillating as Bessel function as calculated in Chap. 2, the transient decays exponentially with a time constant of $1/(2\gamma) = 26.5$ ns. Also, as evident from the result in Fig. 4.3c3, the frequency of the FID signal is always on atomic resonance. The direct measurement of FID from a weak probe pulse clearly shows that FID can be a transient response from the linear propagation. Free-induction decay is a dipole radiation process occurring in coherent atom cloud. It is not a spontaneous emission, even though the optical frequency of FID field can be significantly different from the initial driving probe field.

4.1.2 In Cold Atoms with Large Optical Thickness

To increase the optical depth of the MOT, Wei et al. and Chen et al. [2, 3] used a two-dimensional magneto-optical trap, where the cold atoms are aligned as a cloud in the shape of a cigarette. The experimental setup scheme is similar, as shown in Fig. 4.1. The on-resonance optical depth satisfies $\alpha_0 z \gg 1$. The laser beam with wavelength of 795 nm is directed through a 3 GHz acousto-optic modulator (AOM), and the first-order diffractive output constitutes the weak probe beam

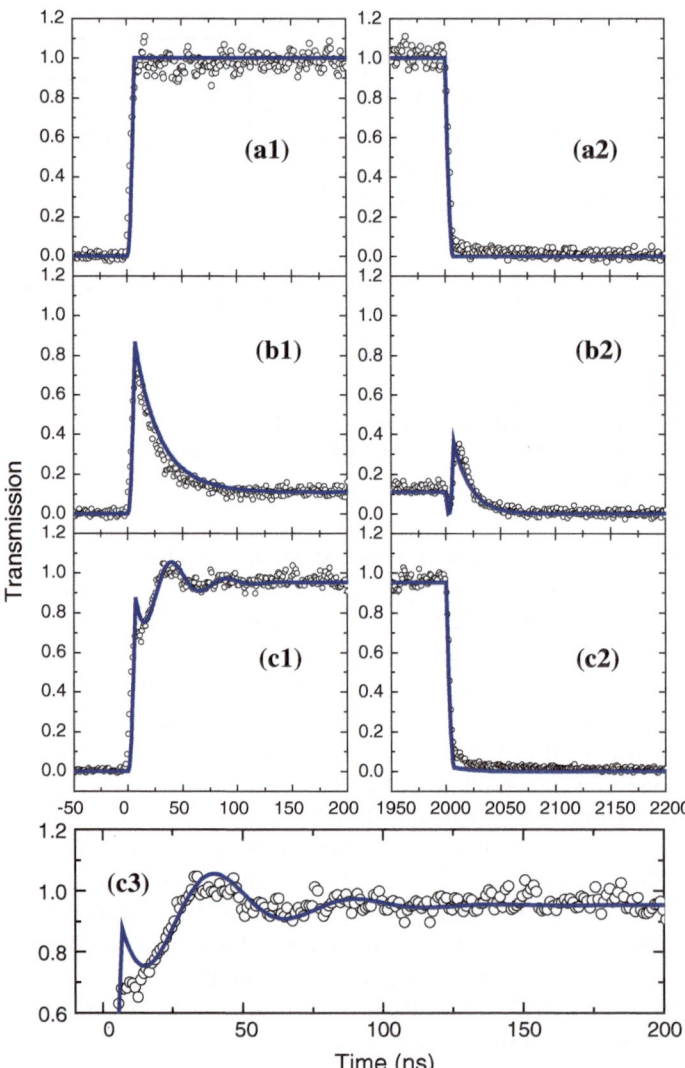

Fig. 4.3 Experimentally observed transient transmission (*black circles*) taken at condition $\alpha_0 z =$ 2.2. **a** The input pulse with sharp rising and falling edges. **b** $\Delta = 0$. **c** $\Delta = 2\pi \times 20$ MHz $\sim 7\gamma$. Panel (**c3**) is an enlarged view of (**c1**). The *blue solid lines* are theoretical *curves* obtained from Eq. (3.15) using FFT. The figure was published in our earlier publication: Chen JF et al. (2010) Phys. Rev. A 81: 033844

whose amplitude is a near-square pulse with a length of 2 μs. With such a sufficiently long square-modulated pulse, optical precursors are displayed both at the rising and falling edges in a single shot.

Figure 4.4 shows the measured optical precursors obtained at the transmitted side after the cold atoms cloud with the on-resonance optical depth $\alpha_0 L = 30$. The input weak near-square pulse, showed in (a) is on-resonance with the atomic transition ($\Delta = 0$). Figure 4.4b shows that the main field is absorbed due to the strong absorption, but the precursors at the rising and falling edges emerge without relative time delay through vacuum. The transmission at the rising and falling edges displays oscillatory structure matching Bessel function. The precursor envelope decays exponentially with a time constant of $1/(2\gamma) = 26.5$ ns. When the carrier frequency of the probe laser is detuned ($\Delta \neq 0$), the spectrum is not symmetric and it is difficult to obtain analytic solutions. Figure 4.5 shows the transients emerging at the rising and falling edges at different probe detuning. The theoretical curves obtained from FFT still match the experimental curves well. As expected, as the frequency of the probe field moves away from the resonant frequency, the leakage of the main pulse becomes dominant in the transmitted pulse and the precursor is diminished by the increasing probe detuning.

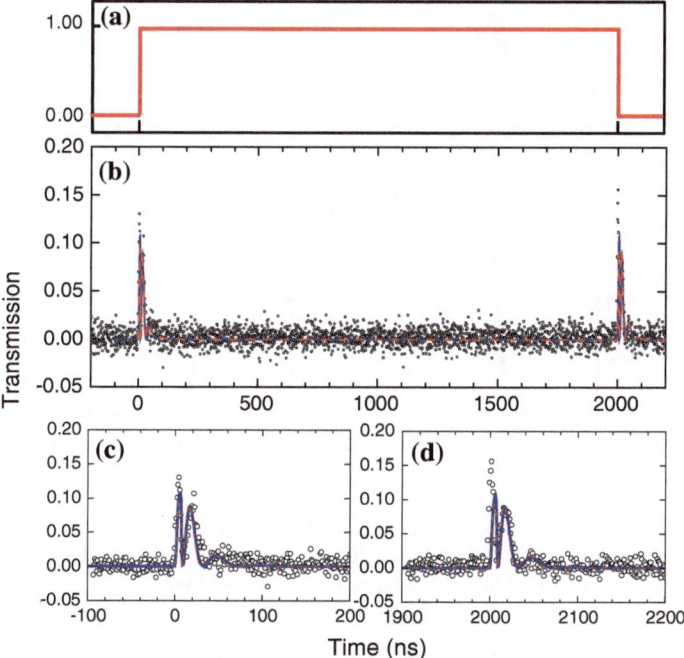

Fig. 4.4   a The input near-square pulse with length of 2 μs; **b** Transmitted signal after the MOT atoms; **c** and **d** specifically are shown the precursors at the rising and falling edge in **b**. The *black circles* are experimental data. The *blue solid lines* are numerical simulation from Eq. (2.14) using FFT. The *red dash lines*, overlapping with the *blue solid lines*, are calculated with asymptotic analysis by taking into account the finite rise and fall time of 7 ns. The figure was originally published in: Chen JF et al. (2010) Phys. Rev. A 81: 033844

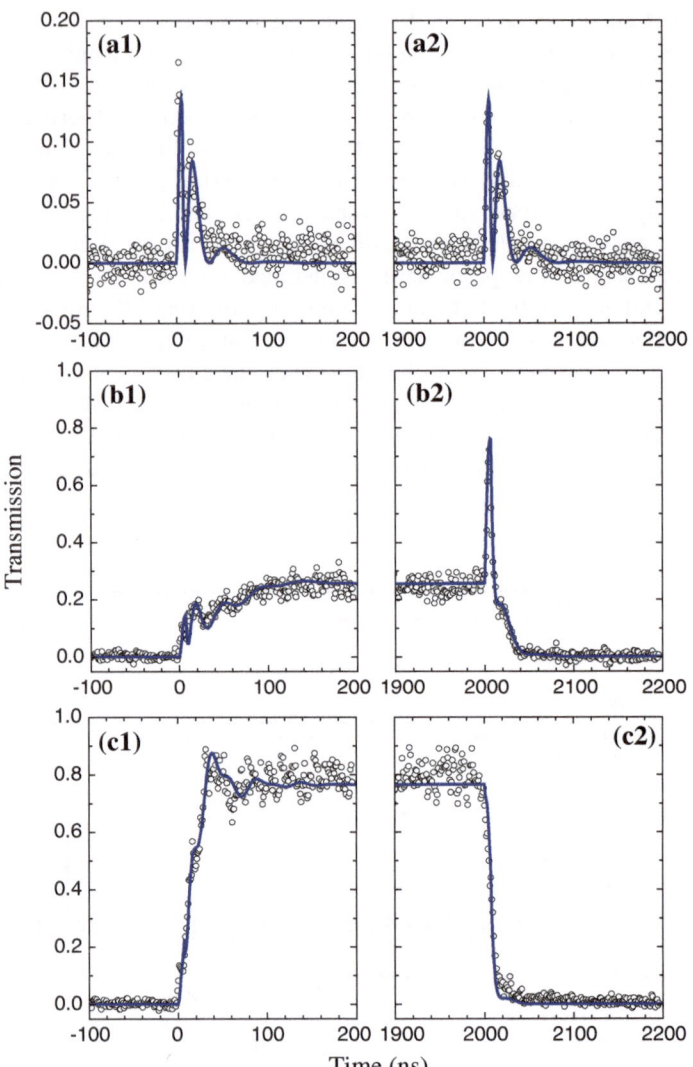

Fig. 4.5 Off-resonance optical precursors measured at the two-level system with $\alpha_0 z = 27$. Rows (**a–c**) list the rising (*column 1*) and falling edges (*column 2*) with $\Delta = 0$, 4γ, 10γ, respectively. The figure is from Chen JF et al. (2010) Phys. Rev. A 81: 033844

4.2 Precursors in a Three-level EIT System

In this section, we review in detail the first observation of precursors in three-level EIT system, which was carried out in a cloud of cold ^{85}Rb atoms. The three-level Λ system is constituted from $|1\rangle = |5S_{1/2}, F = 2\rangle$, $|2\rangle = |5S_{1/2}, F = 3\rangle$ and $|3\rangle =$

$|5P_{1/2}, F = 3\rangle$. The atomic decay rates are $\Gamma_3 = 2\pi \times 6$ MHz, $\Gamma_{31} = \frac{15}{27}\Gamma_3$, and $\Gamma_{32} = \frac{12}{27}\Gamma_3$. Ignoring dephasing caused by collision and thermal motions, the dephasing rates are determined by the lifetime of excited state $|3\rangle$: $\gamma_{13} = \gamma_{23} = \Gamma_3/2$. The dephasing rate of the two ground states is measured to be $\gamma_{12} = 0.01\gamma_{13}$. As Fig. 4.6b indicates, both the coupling and probe laser beams are split from a single laser operated at the ^{85}Rb D1 line (795 nm). Since the probe and coupling beams of the EIT process are from the same laser, the relative phase-frequency noise in the EIT two-photon transition can be eliminated. After a beam splitter, the main laser power passes through an acousto-optic modulator (AOM) and the +1 order serves as the coupling laser that is on resonance with or slightly detuned from the transition $|2\rangle \rightarrow |3\rangle$. The other part of beam power passes through a high-frequency AOM (Brimose) with a central frequency of 3.217 GHz, and the +1 order becomes the probe beam. Therefore, the probe laser detuning can be varied with the AOM central frequency. To make use of most of the multi-Zeeman states for optimizing the EIT effect, both the coupling and probe lasers are identically circularly polarized (σ^+). The coupling beam is almost collinear with the probe beam, but slightly deviated by 2° to avoid entering into a photomultiplier tube (PMT).

The transmitted signal is obtained from the PMT. The included EIT process mainly influences the main pulse which is absorbed completely in the two-level system. Firstly, the EIT transparency window let the resonant spectral components of the input pulse get through the medium, almost without loss, as indicated in Fig. 4.7a where the main pulse transmission is more than 90 %. Secondly, the main pulse transmitted through the EIT transparency window propagates with a slow group velocity $v_g < c$, and hence it is delayed relative to the precursor signal. As shown in (a), the main pulse is switched on after 250 ns (the time when the field grows to 50 % of transmission). The main field is basically composed of

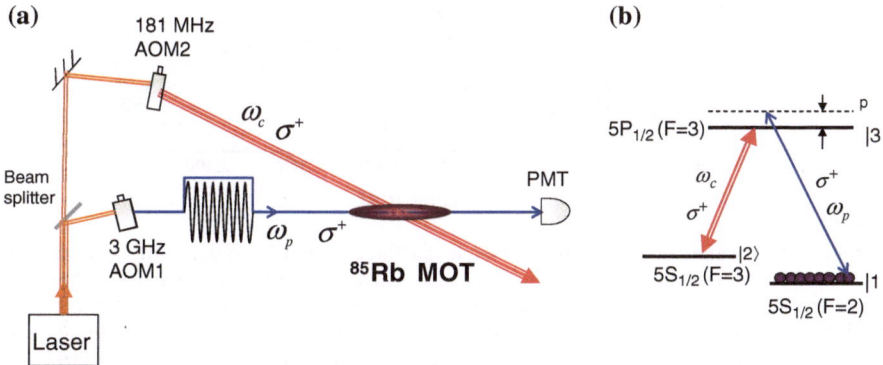

Fig. 4.6 Experimental configuration for optical precursor measurements in three-level Λ system. **a** Measurement scheme. Coupling and probe laser beams are directed from a single laser, and are modulated with individual acousto-optic modulators. **b** Energy level diagram of ^{85}Rb D1 line (795 nm)

resonant spectral components, which severely interacts with the EIT medium and thus are slowed down. In contrast, the precursor field, with far off-resonant spectrum, barely interacts with the medium and thus travels with the maximum speed c. Fig. 4.7b shows that the precursors at the rising edge have the same shape as those observed in the two-level case, and c indicates a different pattern at the falling edge. The precursors generated from the falling edges interfere with the delayed main field. The damped oscillatory precursor field is superimposed onto the transmitted main field and thus the peak transmission is increased to 150 %. One more interesting feature shown in Fig. 4.7 is that, on top of the main field where it is switched on completely, there is an oscillation with a longer period. This is called the postcursor, described by a convolution with an Airy function described in Chap. 2, which was first predicted by Macke and Segard [4].

For completeness, we are also going to show the detuning case for EIT system. Compared with Fig. 4.4, both the two-level and EIT systems give nearly identical results at large probe detuning. This is because the EIT transmission profile is similar to that of the two-level systems at two far-off-resonance wings (Fig. 4.8).

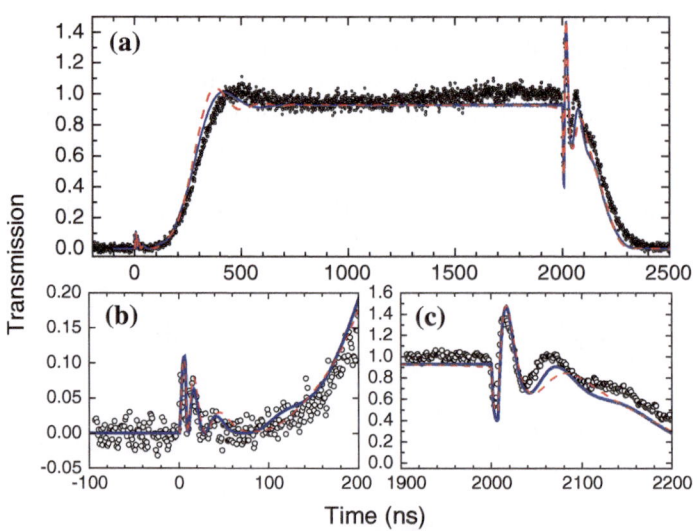

Fig. 4.7 Observation of optical precursors in an three-level EIT system with optical depth $\alpha_0 z = 30$. Coupling field in the EIT process can be characterized by Rabi frequency $\Omega_c = 4\gamma_{13}$. The *black circles* are experimental data. The *blue solid lines* are numerical simulations using FFT. The *red dashed lines* are calculated with hybrid-asymptotic analysis by taking into account the finite rise and fall time of 7 ns. The figure was published in: Chen JF et al. (2010) Phys. Rev. A 81: 033844

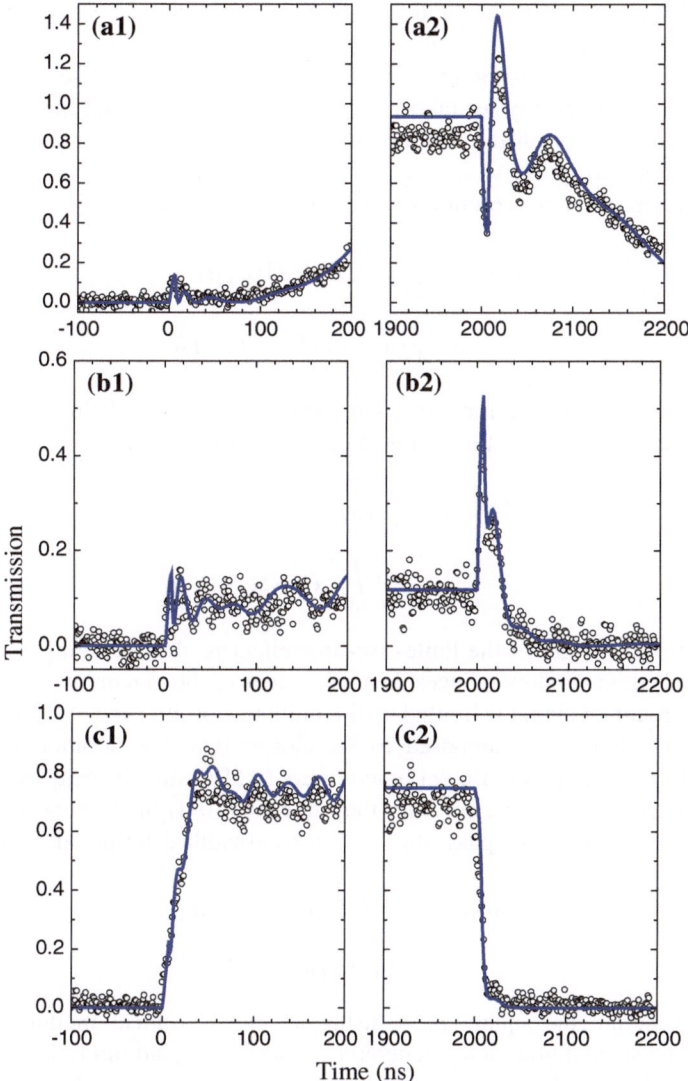

Fig. 4.8 Off-resonance optical precursors measured at the three-level Λ system with $\alpha_0 z = 27$. Rows (**a–c**) list the rising (*column 1*) and falling edges (*column 2*) with $\Delta = 0, 4\gamma, 10\gamma$, respectively. The figure was published in: Chen JF et al. (2010) Phys. Rev. A 81: 033844

4.3   Finite Rise Time and Fall Time Effect on Optical Precursors

In the above measurements of optical precursors, it is obvious that the peak transmission of the precursor signal decreases as optical depth grows. This effect is caused by the finite rise or fall time of the step on or off pulse. Oughstun [5] has

theoretically investigated the finite-rise-time effect on the precursor field formation. The rise and fall times in the previous sessions are taken as 7 ns.

Oughstun suggested using a hyperbolic tangent function to describe a real step
pulse, while for simplicity we take a simple approach and model the realistic step
pulse by turning on or off the field amplitude linearly with a finite rise or fall time
Δt. The real square pulse with length T and rise (fall) time Δt can be mathemat-
ically expressed as a convolution of two ideal square functions [6]:

$$
\begin{aligned}
\tilde{E}_0(t) &= \frac{E_0}{\Delta t} \Pi(t, T) \otimes \Pi(t, \Delta t) \\
&= \frac{1}{\Delta t} \int_0^{\Delta t} E_0 \Pi(t - t', T) dt'
\end{aligned}
\tag{4.1}
$$

in which the unit square function is defined as $\Pi(t, t_d) = 1$ for $0 \leq t \leq t_d$ and
otherwise zero. Therefore, the output precursor field becomes:

$$
\begin{aligned}
\tilde{E}_{SB\pm}(t) &= \frac{1}{\Delta t} E_{SB\pm}(t) \otimes \Pi(t, \Delta t) \\
&= \frac{1}{\Delta t} \int_0^{\Delta t} E_{SB\pm}(t - t') dt'
\end{aligned}
\tag{4.2}
$$

Therefore the origin of the finite-rise-time effect is: the averaging effect within
the rise (fall) time window reduces the peak values of the precursors. The visibility
of optical precursor signal is limited by the finite rise or fall time of the input pulse.
The rise or fall time is supposed to be shorter than the duration of the first
oscillation described by the Bessel function Eq. (2.59), and the decay time constant
$1/(2\gamma)$. Alternatively, if we consider the finite rise (fall) time effect in frequency
domain, it works as a low-pass filter with a bandwidth determined by $1/\Delta t$:

$$
\begin{aligned}
\Phi(\omega) &= \frac{1}{\Delta t} \int \Pi(t, \Delta t) e^{-i\omega t} dt \\
&= \sin c(\omega \Delta t/2) e^{-i\omega \Delta t/2}
\end{aligned}
\tag{4.3}
$$

If we generate the square pulse using an electro-optical modulator (EOM)
driven by the same digital delay generator, a shorter rise and fall time of $\Delta t = 3$ ns
is achieved. The results obtained from the two-level system are shown in Fig. 4.9.
The overall output transmission profiles are the same as the case of $\Delta t = 7$ ns.
However, the normalized peak intensity of the precursor generated from step pulse
with rise time $\Delta t = 3$ ns is about 27 %, which shows a significant increase compared to that of $\Delta t = 7$ ns in (b). With this in mind, we are able to explain why the
precursor signal seems to be heavily diminished in optical thick medium, which
weakly interacts with the precursor field. The saddle points contributing to the
Sommerfeld-Brillouin precursor move far away from the atomic resonance, and
thus are attenuated by the filter effect caused by the finite rise and fall time. The
next section includes more results.

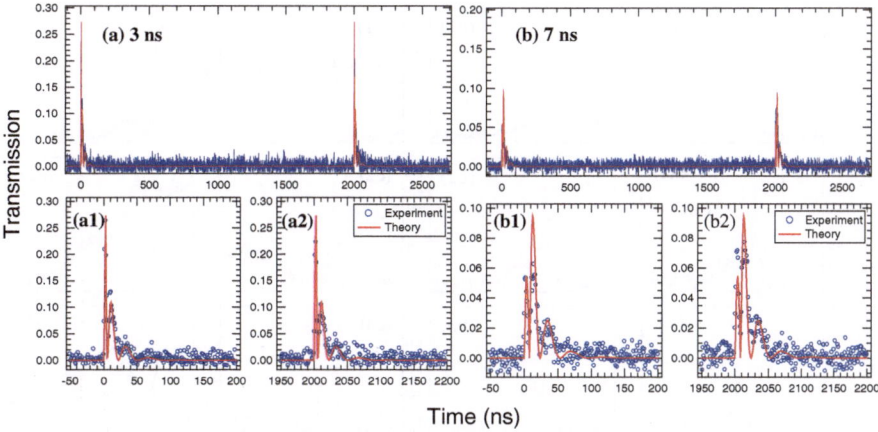

Fig. 4.9 Observation of precursors through a two-level system, $\Omega_c = 0$, $\alpha_0 z = 42$. Two cases of rise and fall times (*3* and *7* ns) are shown and compared. The figure was originally published in: Chen et al. (2010) J. Opt. 12: 104010

4.4 Changing the Optical Thickness of the Medium

As discussed in the previous two sessions, the cold atomic ensemble is converted from an EIT three-level system to a two-level system simply by controlling the coupling laser on or off. Moreover, the optical depth of the atomic cloud can be varied from 0 to 45, by changing the density of the trapped rubidium atoms.

As the optical depth increases, the near-resonance frequency components are strongly absorbed by the medium. Figure 4.10 shows the details of optical transients at the rising and falling edges, while comparing the cases with different optical depths. From $\alpha_0 z > 10$, the transient can be more precisely described in terms of Sommerfeld-Brillouin precursor, exhibiting damped oscillation. At low optical depth of $\alpha_0 z \sim 2$, the transient spike at the rising or falling edge has an exponential decay profile, which is significantly distinguished from the cases of $\alpha_0 z > 10$. With such a low optical thickness, the optical transients should be described as free-induction decay, which stems from the radiation of the macroscopic dipole constituted by the atom cloud. This will be discussed in more detail at the end of this session.

I would like to conclude with some general features for precursor. Firstly, from Fig. 4.10 column a and b, the precursors that emerged at the rising and falling edges are identical. Secondly, at $\alpha_0 z > 10$, one could hardly distinguish between two-level and EIT system. The reason is, at high optical depth condition, the spectral components contributing to the precursor signal move to far-off-resonance regime, where the dispersion curves for two-level and EIT material are identical. On the other hand, Fig. 4.10 indicates that, wave front at the rising edge, or similarly at the falling edge, travels with the speed of light c independent of the atomic density. There are three characteristic times that determine the precursor

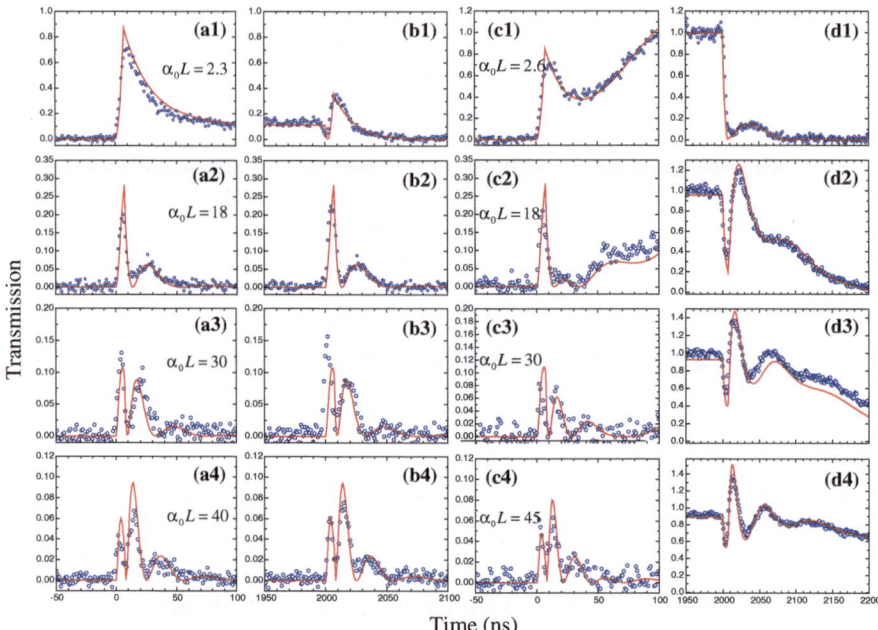

Fig. 4.10 Optical precursors at the rising and falling edge from a two-level system [(**a1**)(**b1**–**a4**)(**b4**)] and three-level EIT system at different optical depth conditions. For the two-level system, the coupling laser is switched off; for the three-level EIT system, the coupling laser is applied with $\Omega_c = 4\gamma_{13}$. The *blue circles* denote the experimental data. The *solid red curves* are obtained from frequency domain linear dispersion approach using FFT. The original figure is from: Wei et al. (2009) Phys. Rev. Letts. 103: 093602

signal. The first is the oscillation peak duration determined by the Bessel function $\tau_s = \pi^2/(2\alpha_0 z\gamma_{13}\tau)$. With increasing optical depth $\alpha_0 z$, the oscillation duration becomes shorter. The second is the precursor (intensity) decay time constant $\tau_\gamma = 1/(2\gamma_{13})$ determined by the atomic natural linewidth. Therefore, for all optical depth cases shown in Fig. 4.10, the decay times $\tau_\gamma = 26.5$ ns are the same for cold rubidium atomic ensemble. Finally, the group delay time τ_g determines the separation between precursor and the main field.

By varying the optical depth from 0 up to 45, we measure the peak transmission of the optical transients for $\Delta t = 3$ and 7 ns, respectively. Also, the theoretical curves calculated for 1 ns rise time are plotted for comparison. Figure 4.11 suggests that higher precursor intensity can be achieved by using a faster light modulator. We find distinct transmission tendencies in the two regimes. Below $\alpha_0 z = 5$, the optical transients grow with increasing optical depth and share the same transmission peak value among different rise (fall) time, while the situations change at high optical depth condition, above $\alpha_0 z = 5$. Figure 4.12 explains the intensity transmission profiles $|H(\omega)|^2$ for both two-level and EIT systems. The field strongly interacts with the atoms at their absorption resonances. For the two-level system, there is one absorption resonance, as shown in Fig. 4.12a. For the EIT system in Fig. 4.12b, there are two absorption dips separated by about Ω_c. At the resonant frequency of the original two-state transition, a transparency window with 100 % transmission occurs. The dashed lines represent the spectrum of a step-modulated pulse with on-resonance carrier frequency. At low optical depth, when the absorption is not significant, the on-resonance frequency components dominate. These spectral components strongly coupled with the atomic system contribute to the FID signal and dominate the optical transient response at low optical depth. At high optical depth when the on-absorption-resonance frequency components are absorbed, optical precursors start to be formed from the lossless far-detuned frequency components that propagate at the speed of light in vacuum c. In this case, the optical transients are dominated by optical precursors. In the two-level system, the main field is absorbed and the precursors "precede" nothing. In the EIT system, the main field lying in the narrow transparency window is delayed due to the slow-light effect. For the two extreme regimes, we have obtained analytic solutions. For the intermediate regime, the optical transients cannot be clearly classified as FID or precursor, and only numerical solutions are obtained.

The transient peaks following the falling edge of the weak square pulse through the two-level system provides us hints of the evolution from FID to precursor, which is depicted in Fig. 4.11. At low optical depth condition, the transient peaks get enhanced with increasing optical depth because of many-atom collective enhancement. Moreover, the FID field has a narrow line width $\gamma_{13}/2\pi = 3$ MHz and thus, is immune to the effect of finite rise and fall time. At a high optical depth condition, we observe optical precursors alternately, since the far-off-resonance frequency components contribute most due to their low absorption. As the optical depth increases, the Sommerfeld-Brillouin saddle points move further away from the atomic resonance and closer to the filter cut-off frequency. As a result, the precursor transmission peak drops. Figure 4.11 shows that, the optical coherent

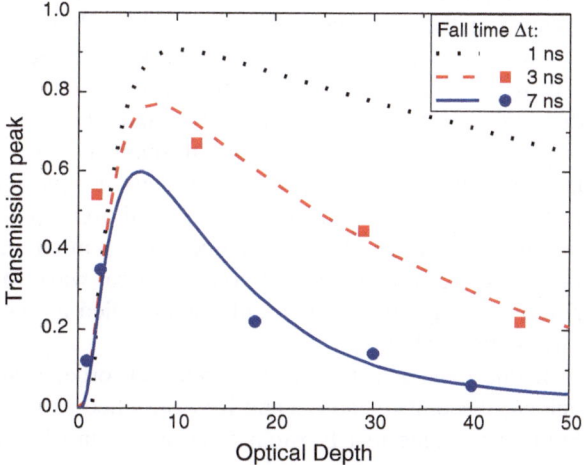

Fig. 4.11 The transmission of the first transient peak after the falling edges as a function of optical depth in the two-level system with the finite fall time of *1*, *3*, and *7* ns. The probe beam in each case is on resonance. The *circular* and *square* makers are experimental data. The *solid* and *dashed lines* are theoretical *curves*. The figure was published in: Chen et al. (2010) Phys. Rev. A 81: 033844

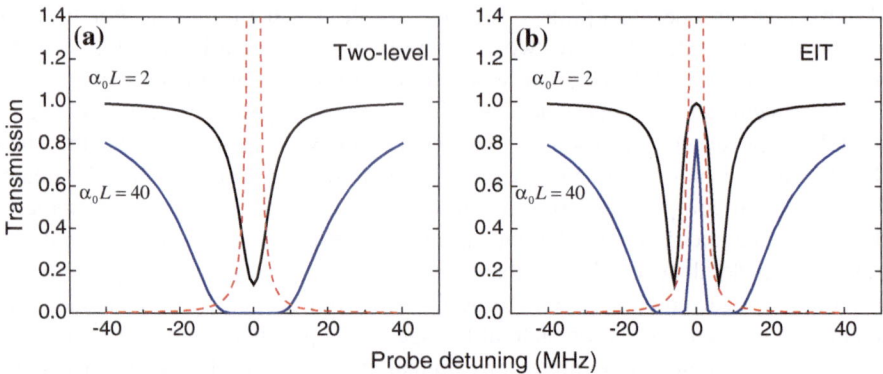

Fig. 4.12 Transmission spectrum of the probe laser in (**a**) a two-level system and (**b**) an EIT system at different optical depths of $\alpha_0 z = 2$ and 40 respectively. The *red dashed lines* denote the spectrum of an input step pulse. The figure was published in Ref. [3]: Chen et al. (2010) Phys. Rev. A 81: 033844

transients excited by a resonant step pulse at low optical depth can be character-ized as FID and those excited by a resonant step pulse at high optical depth as precursors. The transition point is around $\alpha_0 z = 5$.

4.5 Precursors in Superluminal Medium

As more and more physicists are involved in the vigorous discussion of the speed law, which says that nothing can travel faster than the speed of light in vacuum c, fast light propagation in superluminal medium becomes a hot topic. When a Gaussian light pulse travels through the superluminal medium, it is demonstrated that the pulse moves faster than that traveling through vacuum. This is done by comparing the pulse center in both cases. However, the input pulse undergoes severe distortion. A remarkable experiment reported by Wang et al. [7]. showed a tiny peak advance of 62 ns in the superluminal medium, with negligible pulse distortion. In their gain doublet medium, the Gaussian pulse can maintain its original pulse shape when passing through such a gain assisted region. However, how does the information exactly travels in this light pulse? What is the infor-mation velocity v_i? Can it be equal to the group velocity, which is $-c/330$ in Wang's experiment? With these questions in mind, information velocity is defined in the propagation velocity of non-analytical wave front. A sharp step pulse in the rise or fall edge of a square pulse is qualified as such a wave front to clarify the information velocity.

In the case of a two-level Lorentz absorber system, the on-resonance group velocity becomes negative and it is classified as the superluminal medium [8, 9]. Some may argue that Einstein's causality can't be violated in the propagation of step pulse because the main field is essentially completely absorbed at high optical depth. That's why we prepare the two-level system in the low optical depth condition, where the main field is not attenuated to zero. Figure 4.13b shows the propagation of both Gaussian and step pulses at a low optical depth of 2.3 where more than 10 % of the original main field is present. We observe a significant peak advancement of 50 ns in the Gaussian pulse propagation, with obvious attenuation and distortion. However, for the step pulse, we observe no advancement of the rising edge. For comparison, the Gaussian pulse and the step pulse are tested in the EIT system. In the three-level EIT medium, the main field experiences slow light effect and is delayed by the group delay time τ_g. Figure 4.13a shows the propagation of a Gaussian as well as a step pulses through an EIT medium having an optical depth of 30. In the case of Gaussian pulse, we measured a group delay of about 200 ns with no attenuation and distortion. This pulse delay is consistent with the measurement in the case of step pulse case, where the delayed main field turns on smoothly after 200 ns. The leading edge of the precursor shows no detectable delay to the step pulse wave front.

As discussed in Sect. 1.4, both Gaussian and step-modulated optical pulse can be used to encode information, since field amplitudes do change in both cases. However, the advanced Gaussian peak (as shown in Fig. 4.13b) does not represent information propagation in medium. Conversely, the slowly increasing envelope of the Gaussian wave front causes a difficulty for information detection, since it takes a certain amount of time to measure the observable amplitude change. Even when we are able to measure 1/1,000,000 of amplitude change, the time span for such an amplitude change delays the detection of information. The problem of a Gaussian pulse is that it does not have an obvious start point. In contrast, step pulse has a clear start point. However, an ideal step-modulated pulse is not possible in real experiment, and any near-step pulse is characterized by the finite rise or fall time. As long as the rise time is much shorter than the atomic dephasing time, the

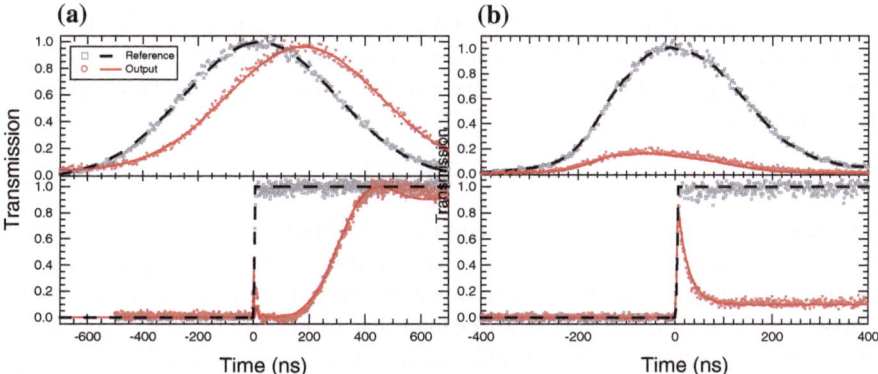

Fig. 4.13 Propagation of Gaussian and step pulses through the medium with slow light or fast light effect. **a** The EIT system with optical depth $\alpha_0 z = 30$ and $\Omega_c = 4\gamma_{13}$; **b** Two-level system with $\alpha_0 z = 2.3$, $\Omega_c = 0$. The figure was published in: Chen et al. (2010) J. Opt. 12: 104010

precursor field is observable and thus be able to flag the information. The precursors observed in the superluminal medium confirm that there is no violation of Einstein's causality principle in light propagation through fast light medium and the information velocity is different from the group velocity.

4.6 Stacked Optical Precursor

Generating optical pulses with high peak power from a low-power laser is of great interest to optical communication, nonlinear spectroscopy, and optical bio-imaging [10, 11]. Therefore, various pulse compression schemes has arisen. A standard pulse compression scheme makes use of frequency chirping followed by a dispersive compensator [12–14]. Alternatively, pulse compression can also be achieved through a nonlinear medium [15–17]. To achieve a short pulse with peak power higher than that of the input beam in the continuous-mode, one can implement frequency-phase modulation to the laser field applied through a near-resonant atomic vapor [16], or pass light through a dispersive modulator [18, 19]. Segard et al. [20] reported peak intensity with three times of enhancement using the electromagnetic pulses in microwave regime. They retained a rotational line $(0 \rightarrow 1)$ of $HC^{15}N$ for excitation, with wavelength $\lambda = 3.5$ mm and linewidth characterized by Doppler broadening of 100 kHz. The molecular gas sample was contained in an oversized circular waveguide (length $l = 182$ m), and the effective optical depth of the sample was 60 approximately. The amplitude of the input pulse was modulated by a series of step pulses and the pulse passes through a resonant absorber. Without experimental demonstration they discussed the possibility of enhancing the peak power with step phase modulation in optical regime [21].

In the narrow resonance cold atomic ensemble described in the former sections, the optical stacked precursors come into experimental observation. The theoretical approach developed in Chap. 2 well predicts the transient behavior. The input pulses with ideal step front generate Sommerfeld-Brillouin precursors. At high optical depth $\alpha_0 z > 10$, the oscillatory precursor field can be approximated as Bessel function:

$$E_{SB\pm}(t) = \pm E_0 J_0(\sqrt{2\alpha_0 z \gamma_{13}(t - z/c)})\Theta(t - z/c)e^{-\gamma_{13}(t-z/c)} \qquad (4.4)$$

in which, $J_0(x)$ is the zeroeth-order first kind Bessel function. If we arrange a series of on and off steps with a time sequence so that the precursor fields produced from all steps at different times interfere constructively, it is possible to generate a transient pulse with higher peak power than that from a single step. Suppose that we design to generate a transient pulse at $t_0 + z/c$, the on and off step sequence applied to the input amplitude can be arranged as [22, 23]:

$$AM(t) = \Theta(t - t_0) + \sum_{i=1}^{N-1} (-1)^{i-1} \Theta(t_0 - t - t_i) \tag{4.5}$$

where N denotes the total number of steps, and $t_i = x_i^2/(2\alpha_0 z \gamma_{13})$. $x_i = (i + 1/4)\pi$ (for $i \geq 1$) is the ith zero of the Bessel function $J_1(x)$ that indicates the position of the extreme of $J_0(x)$ in Eq. (4.4). The peak amplitude of the stacked optical precursor from the amplitude modulation is then obtained at $t_0 + z/c$ as:

$$E_{AM} = E_0 \times \sum_{i=1}^{N-1} (-1)^i J_0(x_i) e^{-x_i^2/(2\alpha_0 z)} \tag{4.6}$$

The above amplitude modulation is not the most efficient scheme to enhance the transient peak because the laser is switched off during some period of dark time. Phase modulation keeps the laser power constant and thus avoids the dark time problem. A maximum phase modulation with sequenced steps in Eqs. (4.4–4.8) can be expressed as:

$$PM(t) = e^{i\pi AM(t)} = -2AM(t) + 1 \tag{4.7}$$

where the second steady-state term does not contribute to the transient response. Therefore, such a phase modulation is equivalent to the amplitude modulation case, but provides a factor of 2 enhancement in the transient field amplitude.

One could further increase the transient spike by adding up the contribution from the main field, which travels through the absorptive material with negligible loss by virtue of the EIT transparency effect [22]. With the on-step arranged closest to the designed point $t_0 + z/c$ as written in Eq. (4.5), we keep the main field in phase with the transient field at time t_0 so that they can interfere constructively. Therefore, instead of Eq. (4.6), the stacked field is expressed as follows:

$$E_{AM} = E_0 \left[1 + \sum_{i=1}^{N-1} (-1)^i J_0(x_i) e^{-x_i^2/(2\alpha_0 z)} \right] \tag{4.8}$$

The experimental setup and energy level scheme is similar with Figs. 4.1 and 4.6. However, after the 3 GHz AOM, an electro-optical modulator (EOM, 10–20 GHz) is inserted into the probe beam path and utilized to modulate the amplitude or phase of the probe beam. An arbitrary function generator generates fast on and fast off step waveform.

In a realistic experiment, the step modulation generated from the modulator has a finite rise and fall time. We know that the finite rise-time effect reduces the precursor transient peak magnitude at high optical depth. However, Eq. (4.6) implies that the peak of the stacked precursor increases as we increase the optical depth because of the decay factor $e^{-x_i^2/(2\alpha_0 z)}$. These two effects will compete and thus the optical depth has an optimum value for generating a high peak power. The peak power generated from the stacked transients from amplitude and phase modulation is optimized with a finite optical depth. In the following part of this section, we would like to describe the recent experiment on stacked precursor observed in cold atoms.

In this particular system, the modulator controlled by the function generator can produce the rise or fall edge with rising time or falling time of 3 ns at best. We found out that the optimized optical depth is in the range between 25 and 35, where the peak power is not sensitive to the change of optical depth. We chose the optical depth $\alpha_0 z = 33$. Due to the time limit of stacking the coherent transient as expressed in the decay term, the terms with $x_i > \sqrt{2\alpha_0 z}$ contribute little to the total transient field. Therefore, for $\alpha_0 z = 33$, we consider the terms with $i \leq 3$ and set the total on–off steps number as $N = 4$, with $\{t_1, t_2, t_3\} = \{12, 40, 83\}$ ns and $t_0 = 1000$ ns. The experimental observations of stacked optical precursors with sequenced-step amplitude and phase modulation are shown in Fig. 4.14a1, b1, c1 show the amplitude modulation case, while a2, b2, c2 show the transients generated from phase modulation. In the two-level system, with amplitude modulation, as Fig. 4.14a1 shows, the steady-state main field is totally absorbed. Because of the finite rise and fall time effect, the precursor peak transmission from a single step is less than 20 % (Fig. 4.6 shows about 10 % transmission for $\alpha_0 z = 30$). The stacking of the succeeding transient peak produces a substantial increase for the

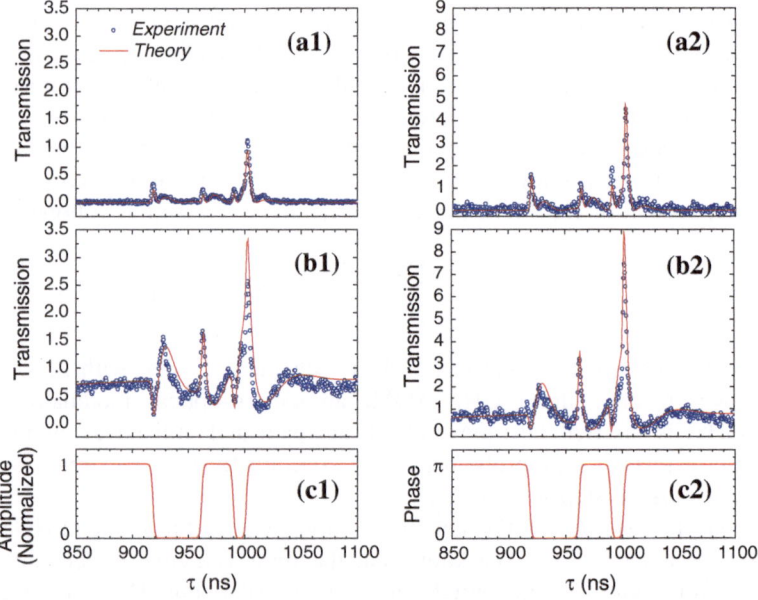

Fig. 4.14 Stacked optical precursor. Column 1 and column 2 depict the cases of amplitude modulation and phase modulation, respectively. The optical depth is $\alpha_0 z = 33$. Transmission of a weak probe pulse with sequenced on–off step modulation is shown in (**a**) two-level system with coupling Rabi frequency $\Omega_c = 0$, and (**b**) EIT system ($\Omega_c = 2.5\gamma_{13}$). Panel (**c1**) is the applied amplitude-modulation waveform, and panel (**c2**) is the applied phase-modulation waveform. The figure is from our earlier publication: Chen et al. (2010) Phys. Rev. Lett. 104: 223602

peak transmission, which is about unity as a1 shown, at the designed location t_0. In the EIT system shown in Fig. 4.14b1, with the coupling laser on, where the main field can be preserved after the medium, the peak power reaches about three times that of the input because of the constructive interference between the stacked precursor and the main field. The red solid lines in the figure are simulated with the transfer function (3.18) and fast-Fourier-transform. In this simulation, we use a more accurate function to mimic the realistic step edges, i.e., the hyperbolic tangent function $\Theta(t) \rightarrow [1 + \tanh(2t/\Delta t - 1)]/2$. The simulated results agree well with the experimental data. Alternatively, the modulation waveform can also be applied to the probe beam phase as shown in Fig. 4.14c2, without changing the amplitude of the probe electric field. For comparison, the phase modulation results are shown in the second column. The enhancement on the transient peaks generated by phase modulation is dramatic, but the shapes are almost identical with the amplitude modulated ones. In the two-level system, the transient peak transmission at t_0 increases from 1.2 to 4.5, four times the enhancement in the intensity profile as predicted by Eq. (4.7). In the EIT system, the transient peak goes up to about eight times the input, quite close to the rough prediction which gives 9 times the enhancement relative to two-level amplitude modulation result (2 times the enhancement of precursor field, added with 1 times of the main field, results in 9 times the enhancement of intensity).

In the former sections, the study of precursors suggested that the transient signal may have potential applications in optical communication through absorptive medium. In this section, we verify that the precursor fields can be stacked constructively, and we could obtain high peak output power. Stacking up the precursor fields can be considered as a scheme of pulse compression, with amplitude or phase modulation applied onto the input square pulse which generates precursor fields.

References

1. Jeong, H., Dawes, A.M.C., Gauthier, D.J.: Direct observation of optical precursors in a region of anomalous dispersion. Phys. Rev. Lett. **96**, 143901 (2006)
2. Wei, D., Chen, J.F., Loy, M.M.T., Wong, G.K.L., Du, S.: Optical precursors with electromagnetically-induced transparency in cold atoms. Phys. Rev. Letts. **103**, 093602 (2009)
3. Chen, J.F., Wang, S., Wei, D., Loy, M.M.T., Wong, G.K.L., Du, S.: Optical coherent transients in cold atoms: From free-induction decay to optical precursors. Phys. Rev. A **81**, 033844 (2010)
4. Macke, B., Segard, B.: Optical precursors in transparent media. Phys. Rev. A. **80**, 011803 (2009)
5. Oughstun, K.E.: Noninstantaneous, finite rise-time effects on the precursor field formation in linear dispersive pulse propagation. J. Opt. Soc. Am. A: **12**, 1715 (1995)
6. Chen, J.F., Loy, M.M.T., Wong, G.K.L., Du, S.: Optical precursors with finite rise and fall time. J. Opt. **12**, 104010 (2010)

7. Wang, L.J., Kuzmich, A., Dogariu, A.: Gain-assisted superluminal light propagation. Nature **406**, 277–279 (2000)
8. Büttiker, M., Washburn, S.: Ado about nothing much. Nature **422**, 271 (2003)
9. Stenner, M.D., Gauthier, D.J., Neifeld, M.A.: The speed of information in a "fast light" optical medium. Nature **425**, 695 (2003)
10. Denk, W., Strickler, J.H., Webb, W.W.: Two-photon laser scanning fluorescence microscopy. Science **248**, 73 (1990)
11. Nakazawa, M., Yamamoto, T., Tamura, K.: 1.28 Tbit/s-70 km OTDM transmission using third- and fourth-order simultaneous dispersion compensation with a phase modulator. Electron. Lett. **36**, 2027 (2000)
12. Treacy, E.B.: Compression of picosecond light sources. Phys. Lett. A **28**, 34 (1968)
13. Grischkowsky, D.: Optical pulse compression. Appl. Phys. Lett. **25**, 566 (1974)
14. Verluise, F., Laude, V., Cheng, Z., Spielmann, C., Tournois, P.: Amplitude and phase control of ultrashort pulses by use of an acousto-optic programmable dispersive filter: pulse compression and shaping. Opt. Lett. **25**, 575 (2000)
15. Nakatsuka, H., Grischkowsky, D., Balant, A.C.: Nonlinear picoseconds-pulse propagation through optical fibers with positive group velocity dispersion. Phys. Rev. Lett. **47**, 910 (1981)
16. Nikolaus, B., Grischkowsky, D.: 12 X pulse compression using optical fibers. Appl. Phys. Lett. **42**, 1 (1983)
17. Jones, D.J., Diddams, S.A., Ranka, J.K., Stentz, A., Windeler, R.S., Hall, J.L., Cundiff, S.T.: Carrier-envelope phase control of femtosecond mode-locked lasers and direct optical frequency synthesis. Science **288**, 635 (2000)
18. Loy, M.M.T.: A dispersive modulator. Appl. Phys. Lett. **26**, 99 (1975)
19. Loy, M.M.T.: The dispersive modulator-A new concept in optical pulse compression. IEEE J. Quantum electron. **13**, 388 (1977)
20. Segard, B., Zemmouri, J., Mache, B.: Generation of electromagnetic pulses by stacking of coherent transients. Europhys. Lett. **4**, 47 (1987)
21. Mache, B., Zemmouri, J., Segard, B.: Compression of phase-switched optical pulses in a resonant absorber. Opt. Commun. **59**, 317 (1986)
22. Jeong, H., Du, S.: Slow-light-induced interference with stacked optical precursors for square input pulses. Opt. Lett. **35**, 124 (2010)
23. Chen, J.F., Jeong, H., Feng, L., Loy, M.M.T., Wong, G.K.L., Du, S.: Stacked optical precursors from amplitude and phase modulations. Phys. Rev. Lett. **104**, 223602 (2010)

Chapter 5
Optical Precursor of a Single Photon

Abstract The optical precursor refers to the propagation of the front of a step optical pulse that always travels at c, the speed of light in vacuum, in any dispersive medium. In particular, it is directly related to the maximum speed of information transmission. However, the classical precursors are entirely based on the propagation of macroscopic electromagnetic (EM) wave, whereas the envisioned applications, particularly in quantum cryptography, involve the interactions of single photons with atoms or molecules quantum mechanically. In this chapter, we review the heralded single photon source and single-photon waveform reshaping. The first direct observation of optical precursors of a single photon was reported for heralded single photons passing through cold atoms. The classical precursor theory applies to the case of single photons. Also, the causality holds for a single photon. This observation is important for understanding the speed limit of quantum information transmission.

5.1 Introduction

As discussed in previous chapters, optical precursor is a good approach to the study of the propagation limit of optical pulses, which indicates that information carried by any classical optical pulse cannot travel faster than the speed of light in vacuum c. In a classical pulse, there are a number of photons, which are either independent with each other (corresponding to thermal light source) or coherent with each other (corresponding to laser source). The group velocity that we talks about actually refers to the velocity of the center of wave train in the classical domain. When keep track of a photon propagating in a medium, we can identify whether it travels at the same group velocity of the whole train, or at a speed faster than any other photons, including even those experiencing little dispersion. It is beyond the current technology to follow the same photon in a bulk of photons, but it is much easier to send a single photon in a single shot and follow it. Two special cases greatly stimulate people's interest. One is the propagation of a single photon

JF Chen et al., *Optical Precursors*, SpringerBriefs in Physics,
DOI: 10.1007/978-981-4451-94-9_5, © The Author(s) 2013

in a slow-light medium, and the other is in a superluminal medium. Eisaman et al. [1] verified the first case by sending heralded single photons through the EIT medium, in which the single-photon pulse whose spectral waveform matches the transparency window travels at a group velocity controlled by the slow-light effect. On the other hand, as mentioned in Chap. 1, Steinberg' work [2] on superluminal medium did not rule out the probability for individual photons to break the speed limit. Even with a small probability of such a single-photon "superluminal channel" [3], people can send information faster than c and break the causality principle to realize the dream of time flying.

In quantum information processing and quantum network, single-photon source is preferable for carrying information to other "quantum processors", due to its weak coupling with the environment and intrinsic quantum nature. Literatures on the potential of single photons in quantum network are fruitful [4–6]. Furthermore, narrowband paired photons have been developed and serve as a satisfactory approach to generating heralded single-photon source, with temporal waveform well controlled. Thanks to all these advanced technologies, we are now able to answer the questions raised at the beginning of this Brief: Can a single photon break the universal speed limit?

5.2 Heralded Single Photon and Waveform Shaping

In quantum mechanics, a Fock state of the light field $|n\rangle$ refers to the eigenstate of the number operator $\hat{n} = \hat{a}^\dagger \hat{a}$, and its eigenvalue n corresponds to a definite photon number. It is easy to produce single-photon level light source, from which the average photon number in a weak laser light pulse is one. However, a light pulse produced from coherent light source, though weak, still satisfies Poisson distribution and constitutes a coherent state but not a Fock state. To generate a single photon in a Fock state is another big subject. Recently, single photons in a Fock state have been demonstrated to be produced from quantum dot [7, 8], single molecule [9, 10], single atom in cavities [11–13], and atomic clouds [13–17]. Yet, experiments reporting the propagation of a single photon are limited mainly because that the single photon wave function is beyond manipulation. The state of art in generating controllable single photon source utilizes narrow-band paired photons produced from a cold atomic ensemble [17–19]. The coherence time of the paired photon correlation function can be extended to 100 ns ~ 1 μs. Therefore the temporal waveform of a single photon is ready to be manipulated by an electro-optic modulator (EOM) applying on the controlling laser beams or on the signal photon beam.

In this section, we will introduce the concept of heralded single photons by mainly discussing about the heralded single photon formed by the two-photon state. As depicted in Fig. 5.1, in the two-photon state, one of the generated photons (photon 1) serves as the trigger and the other highly correlated one (photon 2)

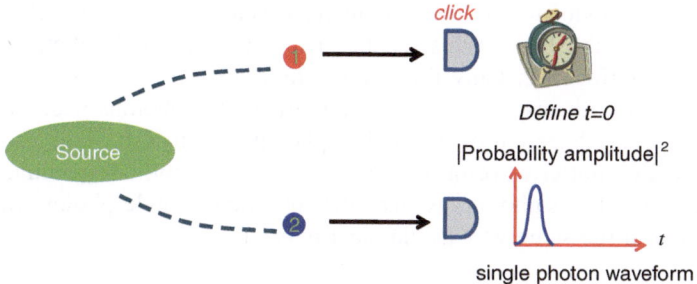

Fig. 5.1 Heralded single photons constituted from correlated photon pairs

therefore constitutes the single photon source. In other words, within the correlation time of the photon pair, photon 2 can be considered as a single photon in a Fock state. The correlation function of the paired photons now serves as the single photon wavefunction, representing the probability amplitude of the arrival of the single photon.

On the other hand, the single-photon quantum nature of the heralded anti-Stokes photons is verified by using a beam splitter (BS), shown in Fig. 5.2. A single photon incident on a BS must be transmitted to port T or reflected to port R, but never both. If the number of the Stokes counts at D_1, the transmitted anti-Stokes collected by D_2, and the reflected beam collected by D_3 are denoted as N_G, N_T, and N_R, respectively, the conditional second-order auto correlation-function is expressed as $g_c^{(2)} = N_{GTR}N_G/N_{GT}N_{GR}$, where N_{GT}, N_{GR}, N_{GTR} are the two-fold and three-fold coincidence counts respectively. A classical field must satisfy $g_c^{(2)} \geq 1$. Convincing proof for the singe-photon nature requires $g_c^{(2)} < 0.5$, since a two-photon Fock state has $g_c^{(2)} = 0.5$.

To reshape the waveform of heralded single photon generated in this manner, Kolchin et al. [20] inserted the EO-modulator in the optical path of one of the paired photons generated via four-wave mixing. The waveform of heralded single

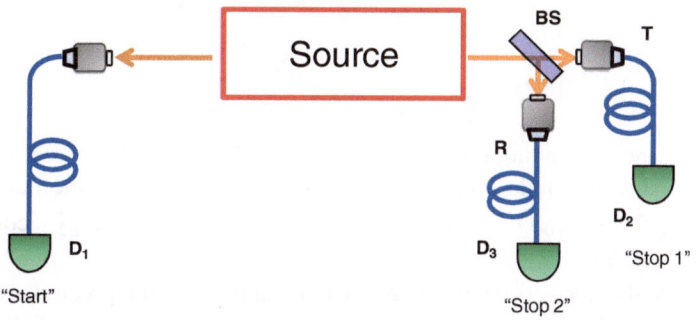

Fig. 5.2 Measurement scheme of conditional second-order auto-correlation function $g_c^{(2)}$

photon may be modulated in the same manner as a classical light pulse, once the time origin is established. In this scheme, reshaping the single photon waveform becomes straightforward. Only the relative time delay ($\tau = t_{as} - t_s$) of the anti-Stokes photons with respect to its counterpart Stokes photons needs to be considered, since the detection of the Stokes photon establish the time origin of the heralded single photon waveform. If we express the amplitude modulation function for the anti-Stokes photons as $m(\tau)$, the conditional single-photon wave packet is modulated in the same way as classical pulse as:

$$
\begin{aligned}
\psi_0(\tau) &= \langle 0 | m(\tau) \hat{a}_{as}(t_s + \tau) \hat{a}_s(t_s) | \Psi_{s,as} \rangle |_{t_s = 0} \\
&= \frac{1}{2\pi} \int \Phi_0(\omega) e^{-i\omega\tau} d\omega
\end{aligned}
\tag{5.1}
$$

where $\Phi_0(\omega)$ denotes the spectrum of the conditional single photon.

5.3 Theory of Precursor of a Single Photon

When the incident optical pulse is generated from a laser beam, one needs to deal with classical electric field in Maxwell's equation in Eq. (2.9). If a single photon is the incident source, an operator representing the quantized electric field should be applied to the Maxwell's equation:

$$
\frac{\partial^2 \hat{E}(z,t)}{\partial z^2} - \frac{1}{c^2} \frac{\partial^2 \hat{E}(z,t)}{\partial t^2} = \frac{1}{\varepsilon_0 c^2} \frac{\partial^2 \hat{P}(z,t)}{\partial t^2}
\tag{5.2}
$$

Similar to the classical electric field, the operator in the time–space can be Fourier transformed into the spectral domain. If we restrict quantized electric fields as electric fields of continuous single mode Gaussian wave in free space, which can be further simplified to plane waves in free space, we obtain the electric field operators for positive and negative-frequency parts as

$$
\hat{E}^{(+)}(z,t) = \frac{1}{\sqrt{2\pi}} \int_0^{+\infty} d\omega \sqrt{\frac{2\hbar\bar{\omega}}{c\varepsilon_0 A}} \hat{a}(\omega) e^{i[k(\omega)z - \omega t]}
\tag{5.3}
$$

$$
\hat{E}^{(-)}(z,t) = \frac{1}{\sqrt{2\pi}} \int_0^{+\infty} d\omega \sqrt{\frac{2\hbar\bar{\omega}}{c\varepsilon_0 A}} \hat{a}^+(\omega) e^{i[k(\omega)z - \omega t]}
\tag{5.4}
$$

where A is the single-mode cross-section area. Considering the real experiment where the incident single photons are emitted into free space, the normalization constant $\sqrt{2\hbar\bar{\omega}/(c\varepsilon_0 A)}$ is determined according to energy conservation $\frac{1}{2} A c \varepsilon_0 E^2 = \langle N \rangle \hbar\omega$.

Therefore the derivations in Chap. 2 can be applied to the precursor of a single photon, except that the electric field now is replaced by electric field operator, which is not observable in experiments. With heralded single photons as source, we could measure the second-order correlation function in the coincidence measurement:

$$
\begin{aligned}
G^{(2)}(t_{as}, t_s) &= \langle \Psi | \hat{a}_s^\dagger(t_s) \hat{a}_{as}^\dagger(t_{as}) \hat{a}_{as}(t_{as}) \hat{a}_s(t_s) | \Psi \rangle \\
&= |\langle 0 | \hat{a}_{as}(t_{as}) \hat{a}_s(t_s) | \Psi \rangle|^2
\end{aligned}
\tag{5.5}
$$

where $|\Psi\rangle$ represents the two-photon state. For convenience we define the two-photon amplitude as $\Psi(t_{as}, t_s) = |\langle 0 | \hat{a}_{as}(t_{as}) \hat{a}_s(t_s) | \Psi \rangle|$. $\hat{a}_{s,as}$ and $\hat{a}_{s,as}^\dagger$ denote the annihilation and creation operator for the paired photons generated (called Stokes and anti-Stokes). By separating a term only relevant to the relative time delay $\tau = t_{as} - t_s$, we can further simplify the two-photon amplitude to $\Psi(t_{as}, t_s) \rightarrow \psi(\tau)$. At this moment, we can view $\psi(\tau)$ as the probability amplitude for a single photon. Note that,

$$
\begin{aligned}
\hat{a}_s(t) &= \frac{1}{\sqrt{2\pi}} \int d\omega \hat{a}_s(\omega) e^{i[k_s(\omega)L/2 - \omega t]} \\
\hat{a}_{as}(t) &= \frac{1}{\sqrt{2\pi}} \int d\omega \hat{a}_{as}(\omega) e^{i[k_{as}(\omega)L/2 - \omega t]}
\end{aligned}
\tag{5.6}
$$

and the commutation relation:

$$
[\hat{a}_s(t), \hat{a}_s^\dagger(t')] = [\hat{a}_{as}(t), \hat{a}_{as}^\dagger(t')] = \delta(t - t')
\tag{5.7}
$$

The incident single photon wave function can be expanded in the frequency domain,

$$
\psi_0(\tau) = \frac{1}{2\pi} \int \Phi_0(\omega) e^{-i\omega\tau} d\omega
\tag{5.8}
$$

where $\Phi_0(\omega)$ represents the spectrum of the conditional single photon. The linear transfer function can describe the single photon with a spectrum of $\Phi_0(\omega)$ passing through the cold atoms in MOT2. The output wave function can be written as,

$$
\psi(\tau) = \frac{1}{2\pi} \int \Phi_0(\omega) e^{i[k_{as}(\omega)L - \omega\tau]} d\omega
\tag{5.9}
$$

Therefore the wave function $\psi(\tau)$ evolves as the electric field amplitude in Chap. 2:

$$
E(z, t) = \frac{1}{\sqrt{2\pi}} \int_{-\infty}^{\infty} E(0, \omega) e^{i\phi(\omega, \theta)} d\omega
\tag{5.10}
$$

In conclusion, the theory in Chap. 2 also applies to the precursor of a single photon. The single photon wave function can be measured through two-photon coincidence as described in the following section.

5.4 Observation of Optical Precursor of a Single Photon [21]

The first part of the system mainly consists of a generation channel for paired photons within the atomic ensemble MOT1. The four energy levels are chosen as $|1\rangle = |5S_{1/2}, F = 2\rangle$, $|2\rangle = |5S_{1/2}, F = 3\rangle$, $|3\rangle = |5P_{1/2}, F = 3\rangle$, and $|4\rangle = |5P_{3/2}, F = 3\rangle$. The atoms are initially prepared in the ground state $|1\rangle$ and pumped to excited states by a pump laser which is blue detuned from $|1\rangle \rightarrow |4\rangle$ transition. Stokes photons are generated from spontaneous Raman transition. The anti-Stokes photons are stimulated by the coupling laser on resonant with $|2\rangle \rightarrow |3\rangle$ transition. Stokes photons are collected by a detector D_1, which triggers both the function generator and the coincidence counts measurement. Anti-Stokes photons are collected by a single mode fiber, before directed into the electro-optical amplitude-modulator controlled by the function generator. Therefore the heralded single photon waveform is modulated with a designed waveform trigged by the detection of the corresponding Stokes photon. The anti-Stokes photons propagate through MOT2 and the precursor can be caught by both detectors D2 and D3. In MOT2, the atomic cloud can be switched between two-level atomic system and a three-level EIT system, by controlling the second coupling beam Ω_{c2}. The parameters in MOT1 are fixed as: $\alpha_0 z = 30$, $\Omega_p = 0.5\gamma_{13}$, $\Omega_{c1} = 3.0\gamma_{13}$, where $\gamma_{13} = 2\pi \times 3$ MHz is the electro dipole relaxation rate between $|1\rangle$ and $|3\rangle$ (Fig. 5.3).

A near-square waveform truncated by the EOM serves as the incident single photon waveform as shown in Fig. 5.4a, with a temporal length of 100 ns and a rise (fall) time of 3 ns. After the heralded single photon (anti-Stokes photons in the generated photon pairs) passes through the EIT system (Fig. 5.4b), the optical precursor at the rising edge is clearly observed. The main wave packet arrives with a 50 ns delay and interferes with the precursor at the falling edge. In the two-level system (Fig. 5.4c), the main wave packet is absorbed whereas precursors at the rising and falling edge remain. Obviously, the precursor and the main signal propagate with different velocity through the dispersive medium. Figure 5.4d plots the signal time delays at different optical depth conditions for MOT2. The main wave packet of the heralded anti-Stokes photon travels at the group velocity whereas the precursor shows no relative delay to the propagation through vacuum.

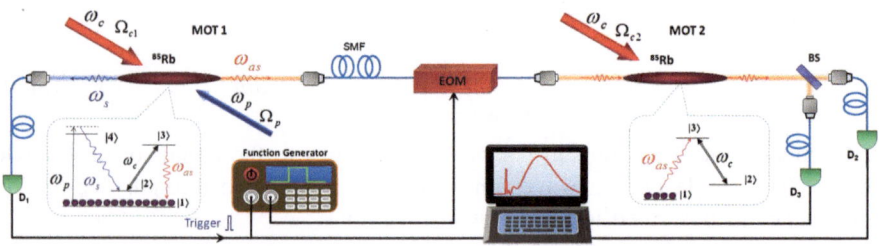

Fig. 5.3 Schematics of the experimental setup. The figure was published in: Zhang [22]

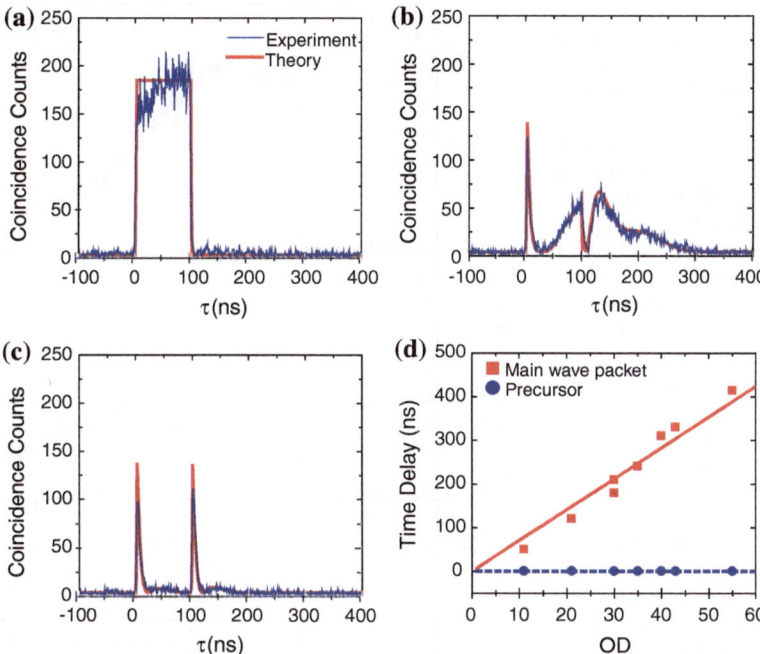

Fig. 5.4 Single-photon optical precursors from a square amplitude modulation. **a** The heralded anti-Stokes photon waveform modulated as a square wave function. **b** and **c** are two-photon coincidence after the anti-Stokes photons passing through the EIT system ($\Omega_{c2} = 3.5\gamma_{13}$, $\alpha_0 z = 10$) and two-level system ($\Omega_{c2} = 0$, $\alpha_0 L = 10$) in MOT2, respectively. **d** The relative time delay (compared with the case of the vacuum) of the precursor and main wave packet as functions of optical depth of the medium MOT2. The *red solid line* is the calculated EIT group delay curve. The coincidence counts are measured with 1 ns time bin. The figure was published in: Zhang [22]

This suggests that the wave front of an optical precursor propagates at c, which is independent of optical depths and other material properties.

We drive the EOM with a step-on waveform as shown in Fig. 5.5a, with a sharp rising edge (the rising time is 3 ns). In comparison, Fig. 5.5b shows the two-photon correlation after the anti-Stokes photon has passed through the EIT medium in MOT2, with $\alpha_0 z = 18$. Similar to Fig. 5.4, the precursor induced by the well-defined wave front is clearly seen at the rising edge and separate from the delayed main wave packet. To investigate the propagation of single photons in a superluminal medium whose group velocity could be negative, we prepare the MOT2 into a two-level system (without the coupling laser) with a low optical depth $\alpha_0 z = 2.5$. In the vicinity of a transition frequency, there exists an anomalous dispersion and therefore the group velocity in this region becomes negative. The negative group velocity in such a fast light medium can be demonstrated by the Gaussian pulse propagation shown in Fig. 5.5d. We observe a peak advance of about 40 ns and a 10 % transmission compared with the propagation through vacuum. Figure 5.5c shows that, for a sharp rising edge, no advancement of any

Fig. 5.5 Single-photon optical precursors from a step amplitude modulation. **a** The heralded anti-Stokes photon waveform with a step modulation. **b** and **c** are two-photon coincidences after the anti-Stokes photons passing through the EIT system ($\Omega_{c2} = 3.8\gamma_{13}$, $\alpha_0 z = 18$) and two-level system ($\Omega_{c2} = 0$, $\alpha_0 z = 2.5$). Inset (**d**) shows the input Gaussian pulse propagation in the two-level system with a peak advancement of about 40 ns (*the lower blue curve*) compared to the reference pulse (*the up green curve*). The *red solid lines* in b, c are calculated from classical wave propagation theory with input square waveform obtained from (**a**). The figure was published in: Zhang [22]

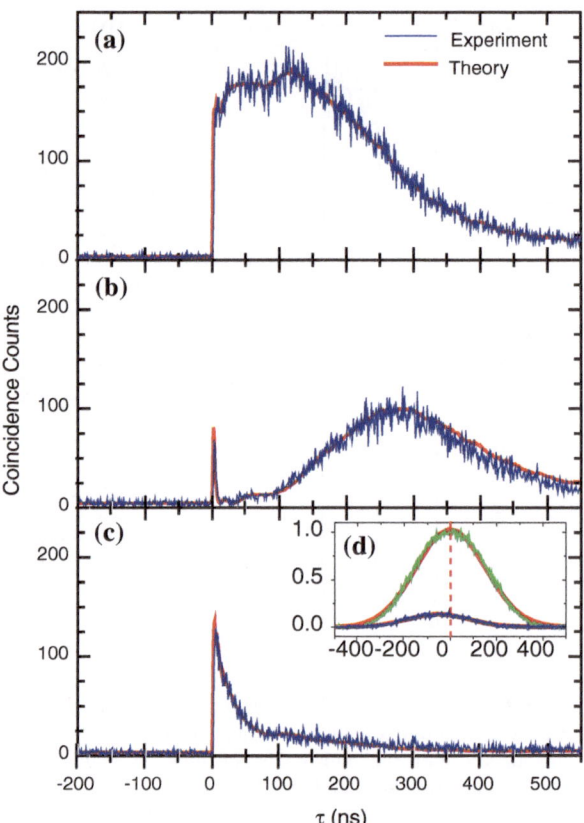

light component can be observed. The well-defined wave front propagates always the in the front of the whole light pulse at a single photon level. The results indicate that the single-photon optical precursor is always the fastest part even in superluminal propagation and Einstein's causality holds for a single photon.

According to Sect. 5.1, $g_c^{(2)} < 0.5$ characterizes single photon source. Taking the measurement shown in Fig. 5.5 for an example, with a coincidence window of 200 ns ($0 \leq \tau \leq 200$ ns), we obtain $g_c^{(2)} = 0.16 \pm 0.11$ before the MOT2. When the heralded single photons travel through MOT2 and the precursor is generated, we obtain $g_c^{(2)} = 0.21 \pm 0.12$. In all these measurements, the precursors are within the coincidence window. The results indicate the near-single-photon characteristics of the heralded anti-Stokes photons and the observed precursors.

The above experimental results also indicate that the optical precursor traveling at c is always the fastest part of the single-photon wave packet in both slow-light and superluminal media. The coincidence counts measurement shows no probability counts for negative delay time, i.e., no counts for single photon with a velocity exceeding c. For the single-photon tunneling, we interpret it as a rearrangement of probability amplitude for a single-photon waveform. As we discuss in the above chapters, information is effectively encoded at the sharp edges of a pulse. Similar to the case of classical light pulses, the sharp rising or falling edges in the single-photon waveform directly determine the start time for sending the single photon. The precursor measurement with single-photon source verifies that, the speed of information carried by a single photon is limited by the speed of light in vacuum.

References

1. Eisaman, M.D., et al.: Electromagnetically induced transparency with tunable single photon pulses. Nature **438**, 837 (2005). (London)
2. Steinberg, A.M., Kwiat, P.G., Chiao, R.Y.: Measurement of the single-photon tunneling time. Phys. Rev. Letts **71**, 708 (1993)
3. Gauthier, D.J., Boyd, R.W.: Fast light slow light and optical precursors: What does it all mean? Photonic Spectra Jan Features **41**(1), 82–92 (2007)
4. Kimble, H.J.: The quantum internet. Nature **453**(7198), 1023–1030 (2008)
5. O'Brien, J.L., Furusawa, A., Vučković, J.: Photonic quantum technologies. Nat. Photonics **3**, 687–695 (2009)
6. Raymer, M.G., Srinvasan, K.: Manipulating the color and shape of single photons. Phys. Today **65**(11), 32–37 (2012)
7. Michler, P., Kiraz, A., Becher, C., Schoenfeld, W.V., Petroff, P.M., Zhang, L., Hu, E., Imamoğlu, A.: A quantum dot single-photon turnstile device. Science **290**, 2282–2285 (2000)
8. Santori, C., Pelton, M., Solomon, G., Dale, Y., Yamamoto, Y.: Triggered single photons from a quantum dot. Phys. Rev. Lett. **81**, 1502–1505 (2001)
9. Brunel, C., Lounis, B., Tamarat, P., Orrit, M.: Triggered source of single photons based on controlled single molecule fluorescence. Phys. Rev. Lett. **83**, 2722–2725 (1999)
10. Lounis, B., Moerner, W.E.: Single photon on demand from a single molecule at room temperature. Nature **407**, 491–493 (2000)
11. Parkins, A.S., Marte, P., Zoller, P., Carnal, O., Kimble, H.J.: Quantum-state mapping between multilevel atoms and cavity light fields. Phys. Rev. A **51**, 1578–1596 (1995)
12. Cirac, J.I., Zoller, P., Kimble, H.J., Mabuchi, H. Quantum state transfer and entanglement distribution among distant nodes in a quantum network. Phys. Rev. Lett. **78**(16), 3221–3224 (1997)
13. Kuhn, A., Hennrich, M., Bondo, T., Rempe, G.: Controlled generation of single photons from a strongly coupled atom-cavity system. Appl. Phys. B **69**, 373–377 (1999)
14. Kuzmich, A., Bowen, W.P., Boozer, A.D., Boca, A., Chou, C.W., Duan, L.M.: Generation of nonclassical photon pairs for scalable quantum communication with atomic ensembles. Nature **423**, 731–734 (2003)
15. van der Wal, C.H., Eisaman, M.D., André, A., Walsworth, R.L., Philips, D.F., Zibrov, A.S., Lukin, M.D.: Atomic memory for correlated photon states. Science **301**, 196–200 (2003)
16. Jiang, W., Han, C., Xue, P., Duan, L.-M., Guo, G.-C.: Nonclassical photon pairs generated from a room-temperature atomic ensemble. Phys. Rev. A **69**, 043819 (2004)
17. Chou, C.W., Polyakov, S.V., Kuzmich, A., Kimble, H.J.: Single photon generation from stored excitation in an atomic ensemble. Phys. Rev. Lett. **92**, 213601 (2004)
18. Balic, V., Braje, D.A., Kolchin, P., Yin, G.Y., Harris, S.E.: Generation of paired photons with controllable waveforms. Phys. Rev. Lett. **94**, 183601 (2005)
19. Kolchin, P., Du, S., Belthangady, C., Yin, G.Y., Harris, S.E.: Generation of narrow-bandwidth paired photons: use of a single driving laser. Phys. Rev. Lett. **97**, 113602 (2006)

20. Du, S., Wen, J., Rubin, M.H., Yin, G.Y.: Four-wave mixing and biphoton generation in a two-level system. Phys. Rev. Lett. **98**, 053601 (2007)
21. Kolchin, P., Belthangady, C., Du, S., Yin, G.Y., Harris, S.E.: Electro-optic modulation of single photons. Phys. Rev. Lett. **101**, 103601 (2008)
22. Zhang, S., Chen, J.F., Liu, C., Loy, M.M.T., Wong, G.K.L., Du, S.: Optical precursor of a single photon. Phys. Rev. Lett. **106**, 243602 (2011)

Chapter 6
Discussion and Outlook

Abstract We discuss the potential applications of optical precursors. Precursor fields are generated from the linear dispersion effect, such that the precursor fields can be stacked to achieve extremely high transient pulse. Another application in pulse manipulation is stimulated by the precursor generated from phase step-modulation, which may be applied to differential phase shifted key scheme. Communication in dense material is another possible application of precursor. Constructed from far off-resonance spectral components regardless of specific medium, precursor finds further advantage in communication under water.

The existence of optical precursor was predicted 100 years ago by Sommerfeld and Brillouin, when they studied the speed limit of a light pulse. Asymptotic analysis was introduced to theoretically calculate the transient signals, and it is found that saddle points contribute to the precursor fields. Many experimental attempts were made, with medium from solid, superfluid, to water. The universal problem, broad-resonance excitation, slowed down the progress of experimental observation. Modern laser cooling and trapping technique solve this problem and offer an ideal medium, cold atoms cloud with narrow resonance, for the optical precursor observation. Sharp rising and falling edges in the optical pulse excite the precursor fields, which propagate with considerable amplitude even in absorptive gas cloud. Precursors are also observed in single-photon propagation, in which the light pulse should not be described by classical electric field, but the probability amplitude from quantum theory. Optical precursors always travel with the speed of light in vaccum c, at the very front of the whole pulse train, either in EIT slow-light medium or superluminal medium. We have to emphasize again that the precursor signal is totally attributed from the linear propagation effect. Very weak optical excitation is enough for measurement. This feature facilitates the future wide application in optical communication in underwater or imaging in biological tissue. In this chapter, we discuss some potential applications.

JF Chen et al., *Optical Precursors*, SpringerBriefs in Physics,
DOI: 10.1007/978-981-4451-94-9_6, © The Author(s) 2013

6.1 Pulse Manipulation

Precursor fields are available to be stacked and constitute higher pulse transients. As discussed in Chap. 5, with the input probe field amplitude or phase modulated by predesigned, sequenced, on–off step waveforms, one could easily obtain stacked optical precursor. It is a linear effect that optical fields constructively build up with each other. Also, a remarkable advantage is that the stacked precursor fields are simply generated from linear passive absorptive material. Such a profound transmission enhancement is a direct application of optical precursor, which was predicted to be a tiny signal in Sommerfeld and Brillouin's theoretical work.

If one can shorten the time of switching, the peak transmission of the transient pulse can be improved further by utilizing an optical thick medium. The power reach an upper limit for a specific number of on–off steps, with an extreme that all the transients stimulated by these steps squeeze to unity at very thick optical absorber. Figure 6.1 shows the peak transmission from phase modulation as a function of the rise time at two different optical depths in the EIT system. The stacked peak power is obtained from on–off steps with $N = 10$. As the rise time reduces to <0.1 ns, the peak transmissions go up to about 15 for $\alpha_0 z = 33$, and 18 for $\alpha_0 z = 66$.

Phase modulation with the input pulse composed of a series of phase shifts differed by 180° produces precursor and stacked precursors. Suppose that we encode phase shifting of π (on-step) as "1", phase shifting of $-\pi$ (off-step) as "0". Arrange these steps at points which match the ith zeros of the Bessel function $J_1(x)$, and thus the precursor field could be well predicted by adding all the precursors together. The

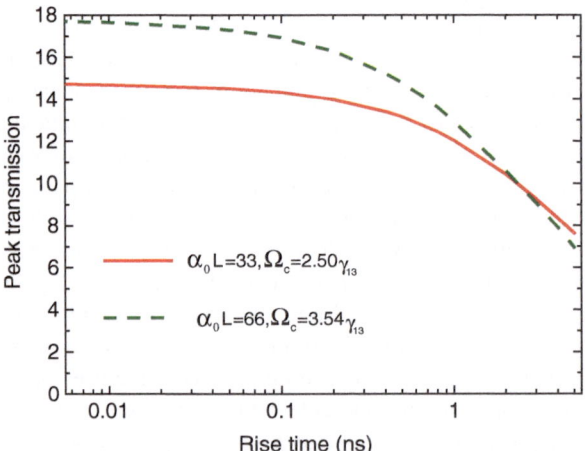

Fig. 6.1 Theoretical *curves* for the precursor peak transmission as a function of the rise time of incident step edges. The *curves* are calculated with EIT configuration, and the applied modulation is phase modulation. For $\alpha_0 z = 66$, the coupling Rabi frequency Ω_c is adjusted to have the same main field transmission (80 %) as that with $\alpha_0 z = 33$. The figure is from our earlier publication: Chen et al. (2010) Phys. Rev. Lett. 104: 223602

size of message carried by one pulse is determined by the available nodes for steps. To maximize the size of message, the demodulator could be consisted of a dense medium. Also, the switching time of phase change is fast enough, for example <0.1 ns. In summary, stacked precursors from phase modulation have potential application in differential phase-shift keying in optical communication.

6.2 Communication

Throughout this book, we emphasize that the speed of information carried by classical light pulse or a single photon is bounded by the speed of light in vacuum c. For now, it is still the fastest speed with which people can send messages to communicate with each other. However, it does not promise that information always propagate with this upper limit, especially when the transmission path is composed of a dispersive medium where pulse reshaping occurs. As discussed in the previous chapters, precursor field is generated from fast switching edge of the incident pulse. The fast changing envelope excites a broad spectrum around the carrier frequency of the incident light, and the extremely high and low frequency components give rise to precursors since these parts of light barely interact with the medium. In other words, a transmitted precursor signal indicates one sudden change of pulse amplitude, no matter switching on or off. In this sense, optical precursor, if observable, is the fastest part of light with meaningful information.

Precursor propagation features low absorption over distance, apart from high speed. Note that the optical depth $\alpha_0 z$ includes distance of propagation z. In theory, the precursor peak always reach unity regardless of the value of $\alpha_0 z$, due to the Bessel function in the precursor field expression. Therefore, increasing z does not decay the precursor peak amplitude. However, as we discussed in Chap. 5, the finite rise and fall time limited by the switching speed of optical fields cause the precursor peak to decay with $\alpha_0 z$. Figure 6.2 shows the tendency with different switching speed of incident pulse. The decay slope of peak amplitude decreases with higher switching speed, or say, smaller rise or fall time. The peak amplitude approaches unity at large propagation distance, provided the switching is fast enough.

In the radio-frequency regime, ultra-wide-band impulse radar arises, and stimulates interest in precursor penetration into foliage. In optical regime, the idea to use precursor for communication under water is being tested continually [1–4], with the help of the femtosecond-pulse technique. To demonstrate why ultra-fast optical pulse can be used to penetrate in dense medium, we simulate the propagation of Gaussian pulse (We express the Gaussian function as $e^{-(t/\sigma)^2}$, with σ as the Gaussian width.) in cold atomic cloud described in Chap. 5, with two different Gaussian pulse width of $\sigma = 0.5$ ns and $\sigma = 100$ ns, respectively. We assume that the carrier frequency of the incident optical pulse is resonant to the transition of two-level atomic system. The results are shown in Fig. 6.3, in which we could compare the propagation of narrow and wide Gaussian pulse. Obviously, the

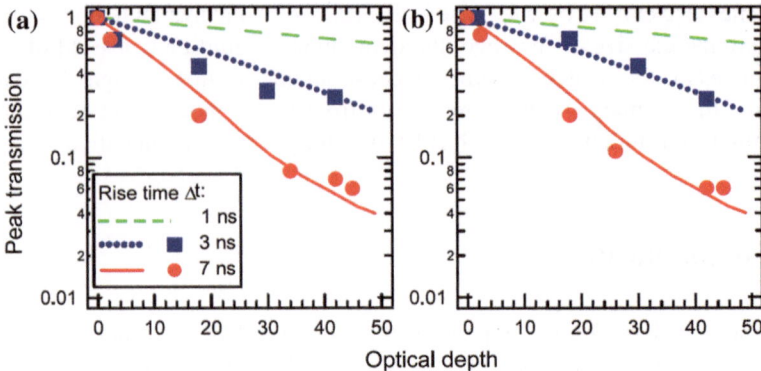

Fig. 6.2 The transient peak transmission versus optical depth. **a** For EIT; **b** for two-level system. The *dots* and *squares* denote experimental data, and *curves* plot the simulation results. The figure was published in J. F. Chen et al. (2010) J. Opt. 12: 104010

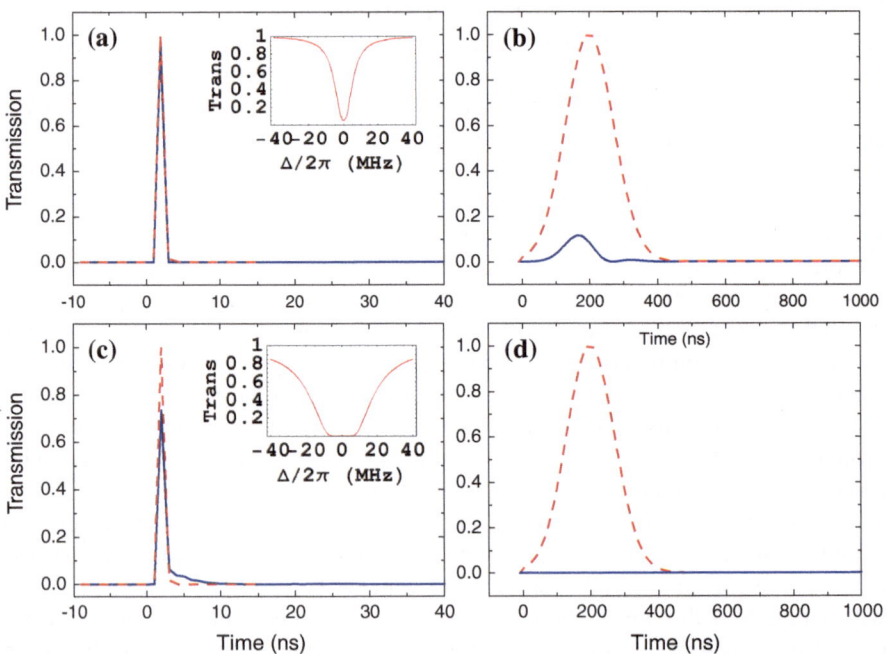

Fig. 6.3 Gaussian pulse propagation in two-level energy structure, with cold atomic ensembles as medium as described in Chap. 5. For (**a**, **b**), OD = 3; for (**c**, **d**), OD = 30. The insets show the corresponding transmission spectrum at each optical depth condition. *Red dashed lines* denote the input pulse, while the *blue solid lines* denote the transmitted pulse through the medium. All the graph are plotted with numerical solutions from Chap. 2

Gaussian pulse with narrow width propagates with little loss even in absorptive medium. The negligible loss is ascribed to the precursor field generated by the ultra-fast pulse, with wide spectrum around carrier frequency. In contrast, the slowly-varying Gaussian pulse is absorbed completely by the medium. Actually, the question whether precursor can improve energy transmission in water attracts quite a few discussions [5] and many attempts have been on-going. It will be a breakthrough if communication under water works without huge construction of fiber system. From the absorption curve, there is a region of single Lorentz function at near infrared regime (700–800 nm) [6]. The FWHM of this absorption curve is around 40 nm, and thus the transient stay for only several picosecond. When future detectors and modulators improve to reach femto-second scale for resolution, infrared optical pulses transmit signal in water efficiently.

Exploring the propagation of precursors in water extend to materials constituted mainly from water, for example, biological tissue [7]. Firstly, the problem was studied in biological tissue due to the concern of health effects raised up by the transient signal. It was verified both in calculation and experiments that the short-rise-time microwave penetrates deeper into tissue than ordinary wave. With the same principle, the precursor can be utilized as signal in living-tissue imaging. This is an under-developed topic which deserves future intensive attentions.

References

1. Choi, S., Österberg, U.L.: Observation of optical precursors in water. Phys. Rev. Lett. **92**, 193903 (2004)
2. Roberts, T.: Comments on observation of optical precursors in water. Phys. Rev. Lett. Phys. Rev. Lett. **93**, 269401 (2004)
3. Alfano, R., Birman, J., Ni, X., Alrubaiee, M., Das, B.B.: Comment on "observation of optical precursors in water". Phys. Rev. Lett. **94**, 239401 (2005)
4. Lukofsky, D., Jeong, H., Bessette, J., Osterberg, U.: Precursors and broadband Beer's law: A discussion on sub-exponential decay of ultrafast pulses in water. PIERS Online **4**(8), 854–858 (2008)
5. Lukofsky, D., Bessette, J., Jeong, H., Garmire, E., Österberg, U.: Can precursors improve the transmission of energy at optical frequencies. J. Mod. Opt. **56**(9), 1083–1090 (2009)
6. Segelstein, D.J.: The complex refractive index of water Master's thesis University of Missouri-Kansas City (1981)
7. Albanese, R., Penn, J., Medina, R.: Short-rise-time microwave pulse propagation through dispersive biological media. J. Opt. Soc. Am. A **6**(9), 1441–1446 (1989)